DON'T
KNOCK
THE
HUSTLE

OTHER BOOKS BY S. CRAIG WATKINS

*The Digital Edge: How Black and Latino
Youth Navigate Digital Inequality*

*The Young and the Digital: What the Migration
to Social-Network Sites, Games, and Anytime,
Anywhere Media Means for Our Future*

*Hip Hop Matters: Politics, Pop Culture, and the
Struggle for the Soul of a Movement*

*Representing: Hip Hop Culture
and the Production of Black Cinema*

DON'T KNOCK THE *HUSTLE*

YOUNG CREATIVES, TECH INGENUITY, AND THE MAKING OF A NEW INNOVATION ECONOMY

S. CRAIG WATKINS

BEACON PRESS · BOSTON

BEACON PRESS
Boston, Massachusetts
www.beacon.org

Beacon Press books
are published under the auspices of
the Unitarian Universalist Association of Congregations.

22 21 20 19   8 7 6 5 4 3 2 1

This book is printed on acid-free paper that meets the uncoated paper
ANSI/NISO specifications for permanence as revised in 1992.

Text design and composition by Kim Arney

Library of Congress Cataloging-in-Publication Data

Names: Watkins, S. Craig (Samuel Craig), author.
Title: Don't knock the hustle : young creatives, tech ingenuity, and the
making of a new innovation economy / S. Craig Watkins.
Description: Boston : Beacon Press, [2019] | Includes bibliographical
references and index.
Identifiers: LCCN 2018053531 (print) | LCCN 2018057244 (ebook) |
ISBN 9780807035313 (ebook) | ISBN 9780807035306 (hardcover : alk. paper)
Subjects: LCSH: Knowledge economy—United States. | Young
businesspeople—United States. | Entrepreneurship—United States. |
Technological innovations—Economic aspects—United States. | Cultural
industries—United States.
Classification: LCC HC110.I55 (ebook) | LCC HC110.I55 W38 2019 (print) |
DDC 338/.0640973—dc23
LC record available at https://lccn.loc.gov/2018053531

To young creatives all over the world,
who are often misunderstood
but are poised to build a brighter future
and better world.

CONTENTS

AUTHOR'S NOTE

Four years ago, a team of graduate students and I set out to answer a series of powerful questions about young people, changes in social norms and technology, and the harsh realities of the post-industrial economy. We began interviewing and observing young people in multiple settings. We conducted more than 150 formal and informal interviews and spent hundreds of hours observing millennials at home, work, and play. We talked to them about life, work, their aspirations, and what it is like to come of age in a world marked by historic change and uncertainty. Our work took us all across the United States and to European cities such as Lisbon and London. What we found is, in many ways, a window into a future marked by unprecedented precarity on the one hand and unprecedented opportunity on the other. What we discovered, I recount in these pages.

RESPECT THE HUSTLE

How a Young Latina's Side Gig
Disrupted American Politics

For more than a year Alexandria Ocasio-Cortez had been waiting for this precise moment: 8:59 p.m., June 26, 2018. That was when the polls in her Democratic Party primary contest against incumbent Joe Crowley in New York's Fourteenth District would start to close and the final votes would be tallied. Ocasio-Cortez had campaigned for ten months to win an election that virtually nobody thought she could win. That morning her staff still did not know where they would hold her watch party. It was yet another sign of what a long shot her campaign was. They finally settled on a billiards hall in the Bronx.

On the way to the watch party Ocasio-Cortez was so nervous that she did something out of character for a twenty-eight-year-old: she turned off her phone, refusing to check any of the polls or social media chatter. "Everybody in the car we were in was so nervous," she said later. "We were just like, 'Don't check it, don't check.'"

Ocasio-Cortez had already convinced herself that even if she lost the election, she and the legion of supporters her campaign had ignited to get involved had already won. In order to force a primary, they'd needed 1,250 signatures. She and her supporters easily exceeded that figure, generating more than 5,000 signatures in the cold and snow of wintertime in New York City to force the Fourteenth District's first primary in fourteen years. On the day of the election, she thanked her supporters via her primary communication platform, Twitter (@Ocasio2018): "No matter who the vote is for, every single vote cast today is ours—because we made this election happen."

The young self-described "girl from the Bronx" was not just challenging Crowley. She was practically taking on New York's entire Democratic Party machine. During the campaign her opponent outspent her thirteen-to-one and received endorsements from New York State power brokers like US senators Chuck Schumer and Kirsten Gillibrand, as well as Governor Andrew Cuomo. "Lots of these folks were mad that I didn't ask for permission to run, that I also was not using the traditional structures of power in New York City to try to run," she said. "In my opinion, if women and gender-expanding people want to run for office, we can't knock on anybody's doors. We have to build our own house."

As her car pulled up to the watch party, she noticed a few reporters rushing into the billiards hall. She did not know what was happening, but she sensed something was going on. When she saw one reporter running—"a big dude," she recalled—Ocasio-Cortez rushed out of the car. "I just started running," she said. "I literally ran and I busted through the doors."

The energy was high inside the billiards hall. A young female reporter from NY1, a local cable news station, grabbed the candidate for a quick interview. As Ocasio-Cortez was talking with her, she looked up at a television monitor. Suddenly, her eyes opened wide, and she let out an uncontrolled scream, covering her mouth with both her hands to conceal what can only be described as equal parts shock and elation. The results were in. She had beaten the incumbent and one of the most connected politicians in New York. In true social media fashion, the video went viral via Twitter, Facebook, and several online news outlets the next day marked with the caption "The Moment You Realize You Just Won."

EXPLAINING THE INEXPLICABLE

The *New York Times* called Ocasio-Cortez's victory "the most significant loss for a Democratic incumbent in more than a decade, and one that will reverberate across the party and the country." The *Washington Post* called it the "defining upset" of the political season. No one saw it coming. As a result, the media quickly had to find a narrative to explain the inexplicable: How had an unheralded twenty-eight-year-old Latina beat the heir apparent to Nancy Pelosi as the leader of the Democratic Party in the US House of Representatives? The media quickly settled on a narrative: demographics as destiny.

New York's Fourteenth District comprises parts of Queens and the Bronx. Nearly half, 49 percent, of the residents in the Queens and Bronx district are Latino. More than 70 percent of residents in the Bronx are people of color. During the campaign, Ocasio-Cortez described her district this way: "New York Fourteen is the source of a lot of the workers in New York's service economy. Uber drivers are all over the city but where do they live? They live in places like Parkchester [her Bronx neighborhood]. This is where waitresses live, where cleaners live. This is a working-class community."

Ocasio-Cortez knows firsthand what it is like to be a service worker. After her father died unexpectedly while she was attending Boston University, her mother started cleaning houses and driving a bus to keep their home from being seized by the bank. The 2008 financial crisis had just hit. Like millions of other families across the US, Ocasio-Cortez's family's financial security was upended. But within a week of her father's death she was back in school, working hard toward her degree and holding down jobs, some of them simultaneously. In 2018 she told the *Intercept*, "I come from a working-class background, so you don't really get a ton of time to mourn." After she graduated college, she started waiting tables to help her mother.

The media spin suggested that Ocasio-Cortez beat Crowley because a district made up of 70 percent persons of color with a median household income of about $53,000 had grown tired of being represented by an out-of-touch white politician. Ocasio-Cortez's victory was described as the inevitable outcome of massive population shifts in the Fourteenth District. Michael Blake, a New York assemblyman who represents the Seventy-Ninth District in the Bronx, summed up the mainstream takeaway regarding Ocasio-Cortez's victory when he told the *New York Times*, "A lot of people of color were excited about a young woman of color. People say demographics are destiny and you can't ignore that reality when looking at the numbers there."

But the demographics-as-destiny narrative is woefully inadequate and inattentive to the complex forces at play. Ocasio-Cortez ran as an intersectional candidate emphasizing her Puerto Rican American heritage, youth, gender, working-class identity, and Democratic socialist principles. She was born in the Bronx, lived in the borough, and embodied the district's hardscrabble way of life. Four months before the election, Ocasio-Cortez was bartending at Flats Fix, a taco bar in Union Square. Her opponent, she reminded voters, lived and sent his kids to private school in Virginia.

It turns out that Ocasio-Cortez won all across the Fourteenth District map. As expected, she performed well in heavily populated Latino neighborhoods and her Bronx neighborhood. Steven Romalewski, director of the Mapping Service at the City University of New York, notes that

Ocasio-Cortez's strongest support came from heavily white and gentrifying neighborhoods in Queens, like Astoria and Sunnyside.

Three days after her election, she offered her own rebuttal of the demographics-as-destiny narrative via Twitter (@Ocasio2018):

> Some folks are saying I won for "demographic" reasons. 1st of all, that's false. We won w/voters of all kinds. 2nd, here's my 1st pair of campaign shoes. I knocked doors until rainwater came through my soles . . . We won bc we out-worked the competition. Period.

The soles of the campaign shoes in the image she posted to Twitter were worn down with holes. In that same tweet, Ocasio-Cortez had one last bit of advice for the press, pundits, critics, and anyone else who credited her win to demographics: "Respect the hustle."

THE RISE OF YOUNG CREATIVES AND THE NEW INNOVATION ECONOMY

Ocasio-Cortez reflects the rise of young creatives—artists, designers, media makers, techies, educators, civic leaders, political activists, social entrepreneurs—who are building a new innovation economy in the face of unprecedented social, technological, and economic change. The new innovation economy is a dynamic sphere of creative, entrepreneurial, and civic activity that expands how we think about innovation in three important ways. First, it expands whom we think of as innovators. Whereas innovation hubs like Silicon Valley tend to be homogeneous—that is, white and male—women, African Americans, and activists in sectors like tech and education are among the principal actors in the new innovation economy.

Second, the new innovation economy expands what is defined as innovation. This economy is embodied by enterprises to, for example, design better STEM learning opportunities for low-opportunity youth, mobilize new modes of political activism through savvy engagement with social media, make independently produced games, or create new forms of television and film that reflect sensibilities traditionally ignored by Hollywood.

Third, in this economy young creatives expand innovation into unconventional spaces. The new innovation economy is active in the underserved neighborhoods of Detroit, despite the largest municipal bankruptcy in US history and a downtown-based gentrification machine that neglects the city's Black residents. Innovation is happening in old buildings that offer cheap rent and plenty of opportunities for young creatives to connect, collaborate, and make things. The new innovation economy is also thriving across digital platforms like YouTube, Twitter, and SoundCloud. These physical and virtual spaces make up what I call "the innovation labs of tomorrow."

Ocasio-Cortez took full advantage of such unconventional spaces during her campaign. Two months before the primary election date, she was still holding fundraisers on Facebook Live to raise money to find space for her campaign staff. Her team used her tiny apartment while also sharing a back room with a livery cab company to conduct her campaign for the US House of Representatives.

Ocasio-Cortez's story is powerful not because it is unique but because it is universal and parallels the story of many young creatives nowadays. Her run for Congress was a classic side hustle. She was pursuing her passion project—political office—with very few traditional resources and alongside a string of gigs that paid her bills.

"I started this race, nine, ten months ago. I was working in education, and I was working in a restaurant, and I started this race out of a paper bag. I had fliers and clipboards, and it really was nonstop knocking doors and talking to the community," Ocasio-Cortez told Joe Scarborough and Mika Brzezinski on MSNBC's *Morning Joe* the day after her primary victory.

The political neophyte had no money or name recognition. She did not come from a political dynasty. She had no paid staff. Ocasio-Cortez did not even have money to run any media advertisements. After bartending shifts, she would attend meetings, small gatherings in the homes of her constituents, and modest fundraising parties as part of her bare-bones campaign.

Like so many young creatives, rather than focus on what she did not have, Ocasio-Cortez focused on what she did have. And that was tenacity, tech savvy, a vibrant social network, and the recognition that people like her have to build the world they want to live in. Faced with an economy in which long-term employment and a secure economic future is less than certain, many young people are electing to pursue a different and more creative entrepreneurial path. Among the young creatives I have met, the goal is not to pursue wealth or celebrity but rather dignity and opportunity. A generation ago, choosing to build your own future would have felt unnecessarily risky, but not for today's young creatives.

This was certainly the case with Ocasio-Cortez. For her and many other young people, the 2016 presidential election was a turning point. She launched a GoFundMe campaign on December 18, 2016, to raise money to drive to Standing Rock to support activists on the ground. The $1,000 she raised was used to offer Standing Rock activists supplies, such as bundles of wood, cots, and subzero sleeping bags. She and two friends hopped in an old Subaru and drove more than 1,600 miles to the heartland. Along the way, they stopped and spoke with people in Ohio and Indiana. They also visited Flint, Michigan, the site of one of the worst water crises in US history. Eventually, they made their way to Standing Rock.

The journey was a personal transformation for Ocasio-Cortez. Each of these states and their respective struggles were unique, but they had something in common: they comprised everyday working-class people who were fighting valiantly just to be treated with dignity in the face of powerful corporate and political interests. Ocasio-Cortez found resolve in the face of daunting circumstances. "I felt like at this point we have nothing to lose. And even in a race that just seemed impossible . . . Even on long odds, that doesn't mean we shouldn't try," she recalled.

Her reflections remind me of the many young creatives I met during the course of my research for this book. In the face of dwindling employment prospects, economic uncertainty, and widening inequality, many have decided to pursue a side hustle, entrepreneurial ambition, or civic endeavor. Many have arrived at a similar conclusion: building their own future is not nearly as risky as it may have once been. Like Ocasio-Cortez, they feel they have nothing to lose.

Similar to others in her generation, especially young Latinos and African Americans, Ocasio-Cortez found that her college degree was not an automatic gateway to upward mobility. African American and Latino college students are much less likely to leave college with a degree than their white and Asian counterparts. And when they do earn a degree, African Americans and Latinos are much more likely to experience underemployment—that is, to be employed in jobs that do not require the college degrees that they earned. This was the reality Ocasio-Cortez faced after graduating from college.

Prior to running for political office, Ocasio-Cortez worked several jobs, primarily in hospitality and education. According to the Bureau of Labor Statistics, millennials like Ocasio-Cortez are much more likely than their older counterparts to work multiple jobs. It has become one of the dominant features of the generation and one of several data points that underscore their precarious economic situation.

In some cases, Ocasio-Cortez took on a job like waitressing or bartending to pay the bills. Her work, in instances like this, was functional. Shortly after graduating college, Ocasio-Cortez tried to launch a children's book publishing start-up. Her work, in instances like this, was aspirational. Her decision to run against a Wall Street–backed incumbent was certainly aspirational. Some even called it delusional.

Critics charged that her youth and political inexperience made her an inadequate candidate. But the exact opposite was true. Her experience as a young working-class Latina made her the perfect candidate for her district and for our times. Her family's personal financial crisis meant that she understood just as well as anybody the struggles that working-class families face to live with dignity in a country marred by an accelerating wealth gap. As a recent college graduate, she understood how financial debt and shrinking economic opportunity jeopardize the future of many young people. The morning after her upset victory, she told one cable news channel, "I understand the pain of working-class Americans, because I have experienced the urgency of this economic moment."

THE TECH INGENUITY OF YOUNG CREATIVES

Ocasio-Cortez's team knew that if they played the game that Crowley was poised to play—PAC money, centrist policies, and focusing on the very small percentage of people who had voted in previous mid-term primary elections—they would lose.

Her campaign highlights many of the hallmark features of the new innovation economy powered by young people in media, technology, design, education, and civic life. It is the ingenuity in the application of technology, the cultivation of pivotal social networks, and the grit to persist and even thrive in the face of limited resources that makes the new innovation economy so promising.

Technology is a prominent feature of the story told in this book. Among other things, technology ignites the new innovation economy by lowering the barrier to entry in historically exclusive sectors (e.g., game design, TV production, politics); by enlarging social networks and audiences; and by providing the tools to bring new ideas, products, and services into the world cheaply and rapidly.

It was no surprise that Ocasio-Cortez's campaign had a digital presence far more dynamic and effective than her fifty-six-year-old opponent's. Her team of mostly twenty-somethings used Twitter, Facebook, and Instagram with skill and nuance to tell her story, talk directly with voters, articulate her progressive politics and policies, design and execute its strategy, and build a vibrant community. There were no seasoned political operatives on the campaign.

Team Ocasio-Cortez's use of social media—strategic, purposeful—resembled that of a lean, hungry start-up. She could not afford a designated campaign office, so her digital team used social media and messaging apps like WhatsApp to meet, share ideas, and coordinate the campaign's strategy. Some of her volunteers were extraordinarily tech savvy and designed apps and analyzed social data to help generate insights about her constituents and the messages they were most likely to respond to.

Young people are constantly criticized as passive, narcissistic, or anti-social for their adoption of technology. But over the years they have used technology to pioneer whole new forms of communication. In fact, their tech ingenuity is often quite strategic and designed to spark new modes of storytelling, community-building, and political action. Tech, in the world of young creatives, offers new ways to engage the world.

YOUNG CREATIVES ARE SOCIAL TOO

It would be a mistake to think that the new innovation economy is powered exclusively by digital capital. The ecosystems developed by young creatives are also powered by dynamic social ties and social networks, because innovation is *emphatically* social. In fact, innovation is built on reciprocity, conversation, collaboration, and support. When most people think of innovation, they likely think of the "lone genius": people such as Bill Gates, Steve Jobs, Larry Page, Sergey Brin, or Mark Zuckerberg. But each of these "geniuses" relied on a social network made up of mentors, collaborators, sponsors, and funders. The new innovation economy that young creatives are building is based on the inventive mobilization of vibrant social networks—people—and the flow of vital social assets—resources. Take, for example, the unique "sharing economy" that young creatives have developed.

Popular notions of the sharing economy mistakenly refer to massive platforms like Uber, Lyft, and Airbnb. In truth, those platforms are premised on financial transactions for delivery of a service like transportation or a place to stay; the parties involved do not share anything of value in the conventional way that we think of sharing. The driver does not give the passenger her car or give the traveler a place to stay beyond an agreed upon amount of money and time. In contrast, young creatives benefit from an innovation economy premised, in part, on the exchange of something of value that is anchored in a social rather than a financial transaction. Young creatives share their time, talents, and expertise with one another, building a culture of giving and receiving that fortifies a fledgling side hustle or entrepreneurial pursuit. I see this repeatedly in my fieldwork.

Like many of the young people that I have met in my research, Ocasio-Cortez may have been poor in terms of financial capital or the economic resources she could tap. But as her campaign gained momentum, she became extraordinarily rich in terms of social capital and the social resources she could tap. As she stated throughout the campaign about her opponents, "They've got money, we've got people."

Thousands of people shared their time and talent to help Ocasio-Cortez's upstart campaign. By the end of the campaign her team of a thousand volunteers had knocked on over 120,000 doors, made over 170,000 phone calls, and sent more than 100,000 text messages. Her vast army of volunteers covered the Fourteenth District like a blanket.

HUMAN CAPITAL AND THE NEW INNOVATION ECONOMY

Young creatives also leverage their social capital to enhance their access to human capital. Ocasio-Cortez did not have the financial resources to produce a campaign video. For much of her campaign she simply went without professional-quality video to help advocate for her election. But through her social media connections, she was able to gain access to people skilled in media production. Human capital—that is, talent and expertise—also powers the new innovation economy. Take, for example, the modest video production company that produced Ocasio-Cortez's campaign video that became a viral sensation.

Naomi Burton and Nick Hayes both worked in corporate media. They met at a political event in Detroit and discovered their mutual interest in telling stories that promoted the values of Democratic socialism. Together, Burton and Hayes decided to leverage their experience in corporate communication into a very different kind of media enterprise and launched a start-up in 2017. Their media company, Means of Production, was a side hustle they labored to make into their main gig.

Hayes told the *Detroit Free Press* that Means of Production wanted to provide services for "unions, the working class, progressive candidates and certain nonprofits." Burton and Hayes decided that if they were going to use the power of media to tell stories to persuade people, why not tell stories that align with their vision of the world? In its profile of Means of Production, *Fast Company* asserts that the company both adopts and subverts corporate-style advertising "to sell Americans on socialism." The name of the company, Means of Production, is a nod to the classic writings of the mastermind behind the most enduring and influential analysis of capitalism, Karl Marx.

Burton discovered Ocasio-Cortez the way most young people discover things these days: through social media. She thought that Ocasio-Cortez's message was bold and unapologetic. "It had a clear leftist vision that I could understand as just a normal person, that I thought other working people could identify with," Burton recalled. She direct-messaged the Ocasio-Cortez campaign via Twitter and pitched the idea of a video. A month later Burton and Hayes were in New York.

The video, like the campaign, reflected the side-hustle ethos common in the new innovation economy. There were no high-priced media consultants or large production crews. Ocasio-Cortez, drawing from her working-class background, wrote the script for her voice-over. Hayes described it as "a very small footprint production. It was just Naomi and myself shooting it and a couple of campaign volunteers." The production team and schedule may have been lean, but the impact of the video would be substantial.

The video, shot in cinéma vérité style, opens with Ocasio-Cortez in her tiny Bronx apartment preparing to hit the campaign trail. "Women like me aren't supposed to run for office," Ocasio-Cortez says. "I wasn't born to a wealthy or powerful family, mother from Puerto Rico, dad from the South Bronx. I was born in a place where your zip code determines your destiny."

In the video we see Ocasio-Cortez visiting her local bodega, talking with everyday people, and riding the subway. "I'm an educator, an organizer, a working-class New Yorker," she says. "I've worked with expectant mothers. I've waited tables and led classrooms." The video portrays a relatable, dignified working-class person who possesses the steely ambition to take on America's widening wealth gap. Just two minutes long, the campaign video was also designed to spread across the social media landscape. And spread it did.

In the first day it generated more than 300,000 views online. By the time the primary date arrived, Ocasio-Cortez's video, simply titled, "The Courage to Change," was the most-watched political campaign video of the season, surpassing more than 2.5 million views. The video, made by what one news outlet called "a ragtag group of filmmakers," generated significantly more buzz and attention for its candidate than the ads her opponent spent millions of dollars to run on New York television and during New York Yankees baseball games.

THE SIDE HUSTLE

Millennials live in a world of contradictions. Ocasio-Cortez's story about struggling against historic currents to make a life with dignity is, at once, autobiographical and generational. That someone as intelligent, hardworking, and talented as Ocasio-Cortez was stuck on the working-class treadmill, grinding every day but not getting ahead, tells you all you need to know about the struggles young people face and why they are drawn to the new innovation economy. In 2015 they became the largest segment in the US workforce, and they currently represent more than one-quarter of the nation's population. They are the most-educated generation in US history, and yet stable and meaningful employment remains elusive. They live in the richest economy the world has ever seen, but they earn less than previous generations of young workers. Adjusted for inflation, the median earnings for full-time workers ages eighteen to thirty-four in 1980 was $36,000. In 2013 the median earnings for that same age group was $34,000.

Most young people struggle with the realities of an economy in which work is itinerant, unfulfilling, or incommensurate with their education or expectations. They are coming of age at a time when many of our notions about work, identity, opportunity, and mobility are undergoing profound change. Long-standing jobs are falling away, and jobs that did not exist two years ago are emerging and changing whole industries. It is an age of instability and opportunity. It is also an age in which, according to a 2016 study by the Equality of Opportunity Project at Stanford University, the standard of living for many millennials will fall below that of their parents. The study found that 92 percent of children born in 1940 went on to earn more than their parents. By comparison, only 50 percent of children born in 1980 earn more than their parents. The study suggests that the promise that each generation will achieve a standard of living higher than the previous generation—the "American Dream"—is in peril. In the economy of tomorrow, more young people will be required to design their own careers and define their own notions of success. This explains, in part, why a rising number of young people are practicing side hustles.

Think of the side hustle as an improvisational and creative assertion of agency in the face of uncertain circumstances. The young creatives whom I profile in the pages that follow and the movement they represent did not invent the side hustle. People have been practicing this way of life for decades, sometimes as a result of opportunity, but often out of necessity. The notion of the side hustle, for example, has been a recurring theme in African American life. Years ago when Lena Horne, a legendary Black entertainer born in 1917, was profiled on the television program *60 Minutes*, the interviewer asked her what her father did for a living. She purportedly replied, "My daddy was a hustler." The interviewer responded, "What does that mean?" Horne politely said, "He did whatever he had to do." Horne's acknowledgment about her father's employment offers some historical perspective on the hustle as a way of life for those on the edge of society.

The Black American film wave of the 1970s—commonly referred to as blaxploitation—was powered by what the British media scholar Eithne Quinn calls the "hustler creative." Melvin Van Peebles and his infamous film *Sweet Sweetback's Baadasssss Song* (1971) epitomizes the era's hustler creative. In the face of near total racial exclusion from the production of film, Van Peebles cobbled together a crew of Black and brown below the line workers, nonprofessional actors ("starring the black community"), and an alternative distribution circuit (pornographic theaters) to make a feature film completely outside the studio system in Hollywood.

Hip-hop culture has also been built on the hustle ethos, from Jay-Z's debut single, "Can't Knock the Hustle," in 1996, and the hit indie film *Hustle and Flow* (2005) to Gucci Mane's signature influence in the rise of "trap rap," a subgenre of hip-hop music that highlights the hustling lifestyle through gritty rhymes and bouncy beats.

While hustling may not be new, the sheer number of people pursuing a side hustle suggests a climate of urgency, especially for the youngest employees in the economy. Aminatou Sow, the cofounder of the popular podcast *Call Your Girlfriend*, told *Time*, "I'm part of this work generation that we really believe in having side hustles, where you go to your job that pays you money where nobody kind of takes you seriously or respects you. And you start doing . . . side projects with friends."

The number of Americans employed in what some economists call alternative work arrangements—the so-called gig economy—is steadily rising. This is a reference to temporary workers, on-call workers, contract workers, gig workers, and freelancers. A 2016 study by the National Bureau of Economic Research found that between 2005 and 2015 the percentage of workers engaged in alternative work arrangements increased from 11 percent to 16 percent. The US Bureau of Labor Statistics reports that contingent workers are twice as likely as noncontingent workers to be under the age of twenty-five. For a surging number of young adults, previously nonstandard forms of work are becoming the standard or simply a way of life.

Don't Knock the Hustle is a glimpse into this extraordinary world.

CHAPTER I

"YOU DON'T NEED A LOT"

The Innovation Labs of Tomorrow

When most people walk into an Apple Store, they usually do so to purchase a device, to get a device that they own serviced, or to play around with one of the devices—iPhone, iPad, MacBook—that have made Apple the richest company in the world. But when New York City–based rapper Prince Harvey walked into an Apple Store one day in 2015, he had a completely different purpose in mind: to finish recording his first album. Harvey did not see the Apple Store most of us see, a retail space. Rather, he saw a studio space.

According to Harvey, the decision to finish his album in the Apple store was the result of bad luck and limited resources. Like a lot of aspiring musicians today, his laptop was his studio. He used the computer to record his lyrics and produce the beats that accompanied them. The computer was also where he stored all of his audio files. In the world of do-it-yourself media, personal technologies have become the primary platform for creative expression and entrepreneurship. But a fateful string of events forced Harvey to devise a plan B.

One day while working on the album, Harvey's laptop crashed. Shortly after that, the external hard drive that he used as a backup was stolen. "It wasn't my plan to record this at the Apple Store," Harvey told the *Daily Beast*. "New York is expensive. I couldn't just buy another laptop."

A friend suggested that he use a display laptop in a computer store to finish production on the album. It was a bizarre idea, but Harvey realized he had little choice if he was ever going to finish the album. One morning Harvey hopped on the J train, which took him from Brooklyn to the Apple

Store in Manhattan's SoHo district. This became his routine for the next four months. The Apple Store offered him access to a laptop, music creation software, a built-in microphone, and a pair of headphones. He also used the store's Wi-Fi to email himself some of the audio files that he created.

The twenty-five-year-old rapper spent so much time in the store that "people would also ask me for tips on GarageBand—after a while I almost felt like a surrogate employee there." Harvey tried to blend in with casual dress and a careful monitoring of his lyrics. "Sometimes I would get a little loud—a lot of my music has profanity, so there were some stares."

With no laptop, recording studio, or production team, Harvey was forced to mine the resources he could command—namely, grit, tech savvy, and the desire to be heard. He needed access to technology to produce the album. But he relied just as much on his social and musical ingenuity as he did his tech ingenuity. He basically turned the Apple Store into his very own innovation lab. Harvey also turned his voice into an instrument, using a combination of lyrics, vocals, and beatboxing to provide the "music" and ambience for his tracks. He called the effort "a cappella" hip hop.

Like so many other young creatives, Harvey constantly struggled to maintain personal and financial stability in a city that was becoming increasingly difficult for striving artists to afford. His experience in the Apple Store is a jarring reminder of the class divide in America and a lack of generational wealth for many young creatives.

Rather than use his dismal circumstances as an excuse for not pursuing his art, Harvey persisted. "A lot of people give you excuses as to why they can't get shit done, and why they can't complete their projects or do what they need to do," Harvey told the digital news source *Mic*. "I didn't want to be one of those people. And I also wanted to show my [artist] friends 'Hey, you don't need a lot.'"

Harvey's story went viral, and several press outlets including the *Daily Beast, Inc., BuzzFeed*, and even *Billboard*, the music industry's main trade publication, reported in amazement on his ingenious decision to turn an Apple Store into a recording studio. But the image of Harvey using the Apple Store as a space to perform creative labor resonates with what my fieldwork reveals about young creatives and the new innovation economy. The innovation economy that young creatives are building is as much about social ingenuity as it is tech ingenuity—that is, their ability to use what little they have to pursue what they want, whether that means a new career, creative expression, or social justice. Harvey's story is revealing not simply because he used technology in an inventive way. His story is revealing because he used space and his creative instincts in inventive ways too. His genius was not simply that he used a display laptop to record his album but that he had the audacity to leverage a resource—the Apple Store—in a way that most would have never dared to imagine.

In the face of enormous odds young creatives are using social ingenuity to turn a variety of unlikely venues into spaces to pursue their side hustles and entrepreneurial ambitions. In doing so, they are expanding the geography of innovation.

WHERE THE FUTURE IS BUILT

When you think of the geography of innovation, certain settings likely come to mind. You might think of tech giants like Google and Apple as ground zero for the creation of some of the most innovative ideas, products, and services the world has ever seen. Google's X lab, referred to internally as "the moonshot factory," is a highly secretive space where designers, artists, and engineers benefit from the company's deep pockets to experiment with all kinds of new ideas and products, from contact lenses that measure glucose levels in a diabetic patient's tears to autonomous driving cars. Apple's design lab—where it has created the iPhone and the Apple Watch—is so secretive that only a minuscule number of its employees have ever walked through its doors. In my fieldwork I visited numerous tech hubs in cities from Austin to Boston. In places like these, incubators and accelerators emerge as principal sites for innovation. As a university professor, I am intimately familiar with a host of spaces—classrooms, seminars, research labs, design studios, and makerspaces—that offer a setting and plentiful resources to faculty and students who are heavily invested in the innovation enterprise.

At their best, innovation labs encourage risk-taking, a key element of the invention-failure-iteration method widely practiced by entrepreneurs. The people fortunate enough to have access to these spaces are granted a license to make the unthinkable tangible. These spaces are predicated on the idea of mobilizing valuable sources of capital—financial, social, human, technological—for ideation, experimentation, and innovation. Facebook's street address, 1 Hacker Way, is a testament to the "fail fast, break things" ethos that has been a central principle in the innovation culture that spread throughout Silicon Valley and beyond. Good innovation labs are places to imagine ideas, products, and services and then proceed to make them. Good innovation labs are where the future is built.

Google and Apple are highly exclusive and represent the twenty-first-century power elite—the corporations, venture capitalists, elected officials, and university personnel—that drive the formal innovation economy. The tech companies are powered, of course, by their substantial wealth; the ability to recruit talent in tech, design, and engineering; and their ascendancy in the global economy. Tech hubs benefit greatly from the nexus between venture capital, the business elite, and public officials interested in economic development. Dedicated laboratories for catalyzing research, talent, and federal funding agencies enrich innovators and entrepreneurs working in the university setting. Needless to say, most young creatives do not have access to the physical assets offered by big tech companies, downtown tech hubs, or universities. But that has not stopped them from creating their own spaces in which to tinker, try new things, and experiment with their own version of the future.

The young creatives that we observed have turned to alternative spaces to pursue their entrepreneurial ambitions. Through determination and improvisation they are building what I like to call the "innovation labs of tomorrow." Among other things, these spaces enable young creatives to meet, connect, and share ideas. The innovation labs of tomorrow also offer the opportunity to test new ideas, products, and services. And like all good innovation labs, these spaces afford young creatives the opportunity to experiment and learn new things. Unlike formal innovation labs like Google X, the innovation labs of tomorrow are accessible to a broader segment of the population and rely on alternative forms of capital. By Silicon Valley or university standards, they are notably underresourced. Prince Harvey's story beautifully illustrates how young creatives are accessing new kinds of spaces that reinvent the innovation enterprise into something that is resilient, relevant, and, at times, urgent.

If there is one criticism of millennials that persists in public discourse, it is the idea that their attachment to technology has made them less social and less connected to others. From this perspective, social media is an oxymoron, insofar as it makes people less rather than more social. This idea feeds the widely popular but wrongheaded view that young people do not invest in their communities and the people around them. Moreover, this view suggests that young people are what we might call social-capital poor, that is, bereft of the social ties and networks that are crucial to getting ahead in the game of life or building strong communities. Compared to their older counterparts, most young people do lack a strong and diverse social network. But that is because they are young and just getting started in life, not because they are antisocial. Over the course of our fieldwork, we noticed the many different ways in which young creatives leverage the resources available to them to build an innovation economy that is premised on community-building behaviors like sharing, reciprocity, and collaboration.

As our fieldwork evolved, we were especially struck by the degree to which young creatives actively seek out spaces to cultivate social ties that expand their social network and their ability to pursue their entrepreneurial and civic aspirations. Many of the spaces that young creatives frequent in pursuit of their entrepreneurial ambitions are by nature social. Innovation labs of tomorrow's activities like meetups, workshops, boot camps, hackathons, dinners, and classes serve as spaces that foster conversation, the exchange of ideas, and social interaction. Over time it became clear to us that young creatives understand that innovation is an inherently social enterprise, one that requires a rich social network of reciprocity and support. The innovation labs of tomorrow are a resounding rejection of the commonly held view that members of this generation have retreated from their peers, their communities, and the world around them.

WHY YOUNG CREATIVES LOVE OLD BUILDINGS

The innovation labs of tomorrow include a unique mixture of spaces that attract young creatives looking to remake their lives and economic opportunities. These spaces have four core features in common: they encourage sociability and experimentation, and they are flexible and affordable.

First, several of the spaces that we visited during our fieldwork are what I call social capital hubs. In other words, the spaces that young creatives select for their side hustles enhance their access to material, intellectual, and emotional resources. Second, all good innovation labs invite experimentation and risk-taking. Virtually every space that we observed in our fieldwork afforded young creatives opportunities to create and test new ideas, build prototypes, get feedback, and iterate. These features make them vibrant learning laboratories in which to develop human capital.

One of the chief characteristics of young creatives is their extraordinary flexibility in the face of economic uncertainty. Just when it seemed that Prince Harvey's technology options had been erased, he adapted on the fly and improvised other ways to work on his music. Flexibility is also a hallmark feature of the spaces where young creatives pursue a side hustle or practice innovation. Many of the spaces that we observed were capable of meeting a wide array of demands. Take, for example, an old house that had been converted into a school that offered a nine-month curriculum in interaction design. The house was, first and foremost, a school for young professionals looking to upgrade their job skills or, as I discuss later in this chapter, make a career pivot. But the house was also used for social networking events, start-up weekends, a boot camp for female entrepreneurs, workshops for creative types, a demo space, and office hours that invited young creatives to meet with and share conversations with established entrepreneurs.

Finally, all of the spaces we observed were affordable. In fact, they were almost always free or cost very little to use. This is obviously crucial for young creatives who are long on ambition, deep on grit, but short on cash. In addition to being affordable, many of these spaces were also accessible. The side-hustling economy is usually a nighttime economy, hence the spaces frequented by young creatives need to be available during the times of day— late at night and on the weekends—that serve as some of the prime hours to work and connect in the new innovation economy.

Importantly, the innovation labs of tomorrow are not strictly digital and impersonal; they are physical and extraordinarily social. Rather than retreating from their peers and the world around them, young people congregate in certain physical spaces to invest in networking, reciprocity, and collaboration.

Additionally, the two physical spaces described in this chapter—a coworking space and an event space—evoke the ideas about city planning and urban enterprise that urban activist and author Jane Jacobs developed more than five decades ago. In her most well-known book, *The Death and Life of Great American Cities*, Jacobs is an advocate for cities that are lively, diverse, and open to a variety of creative and entrepreneurial visions. She asserted that the people who were most likely to generate new ideas and innovative enterprises were those who could least afford the high costs of new construction. Jacobs believed that it was the upstart entrepreneur, the artist who thinks outside the box, or the savvy citizen activist who possessed the vision and the disposition to see beyond the status quo. Jacobs writes, "There is no leeway for chancy trial, error and experimentation in the high-overhead economy of new construction." Jacobs famously proclaimed, "New ideas need old buildings." Jacobs's observations are a fitting testament to the new innovation economy that young creatives are building.

Jacobs believed that the decisions made by power brokers—city officials, planners, and architects—undermine the vitality and diversity of urban streets, neighborhoods, and enterprise. Today, her concerns resonate in a formal innovation economy shaped by a new cast of power brokers: tech bros, elite venture capitalists, exclusive universities, and well-connected city officials. The formal innovation economy—think Silicon Valley or Dumbo Brooklyn—is marked by racial and gender homogeneity, exclusion, concentration of wealth, and maintenance of the status quo. Today's young creatives seek to build an innovation economy that is marked by diversity, inclusion, equity, and a rejection of the status quo. The adaptation of old, underutilized, and virtual spaces is a crucial component in the making of the new innovation economy.

THE RISE OF COWORKING SPACES

Americans are in the midst of a stunning remake of the labor force. No greater evidence of this exists than in what economists call the shift from the standard workforce to the nonstandard workforce. This is a reference to the increasing number of individuals who are self-employed or who work, for example, as freelancers, gig workers, and independent contractors. According to a report produced by the National Bureau of Economic Research, the percentage of workers engaged in nonstandard—or what they call "alternative"—work arrangements rose from 11 percent in 2005 to 16 percent in 2015. The report estimates that the nonstandard workforce will continue to grow. This emerging occupational landscape has been decades in the making and represents the shift from a goods-producing to a service-producing economy.

One manifestation of the rise of the nonstandard workforce is the growing popularity of coworking spaces, which are booming across the US and the world. They are both a physical and cultural manifestation of the rapidly changing economy and shifting patterns in work. Members of coworking spaces share things like furniture, printers, phones, Wi-Fi, operational and technical support, conference rooms, receptionists, and coffee and break areas. The number of coworking spaces in 2007 worldwide was fourteen, according to the Global Coworking Unconference Conference. By 2022 that figure is expected to grow to more than thirty thousand. And, after looking at several forecasts, the Coworking Unconference predicts that the number of coworkers around the world will grow from 522,600 in 2015 to roughly five million by 2022.

Visit a coworking space, and you can usually sense the vibe. Some users of coworking spaces value productivity and may be drawn to the amenities typically found in a traditional office, such as Wi-Fi, a meeting room, copy machine, and printer. Other users value community and may be drawn to social properties, such as people to talk to and connect with, and social activities such as dining together or attending happy hour. In the latter iteration, the idea is to not only find a place to work but also find people to connect to.

Over the course of my fieldwork I visited several coworking spaces across the US, including in Boston, Austin, New York, and Chicago. I even visited coworking spaces in European cities like Lisbon and London. They were each unique, and I saw many different approaches to coworking. Several of the spaces were occupied by young people looking to make their mark in tech or social innovation. I met employees of lean start-ups as well as individual contract workers, freelancers, and people just setting out on new career paths. In the sites that I visited, there was a sense that coworking was not only a distinct mode of work but that it was also a distinct mission and pathway to a more promising future.

Researchers are identifying some notable trends that help to explain the formation of coworking spaces and their role in our society and economy. One early study from the University of Texas identified three types of coworking environments: "community work spaces," "unoffices," and "federated work spaces." Community workspaces are defined in the study as mixed-used buildings that do not feature coworking as the primary purpose. Unoffices are coworking spaces where people go to feel as if they are working in an office, parallel with others. And then there are federated workspaces that aim to foster more active connections between coworkers.

Researchers from the University of Michigan's Ross School of Business identified some of the properties that enable users of coworking spaces to thrive. After conducting interviews with coworking space founders, managers, and workers, they identified three predictors for thriving: First, people who use coworking space found their work to be more meaningful. For instance, many believed that they could bring their whole selves to work while also being immersed in a space that was generally supportive rather than competitive. Second, many coworkers enjoyed the feeling of control. In addition to choosing their work, users chose the type of coworking space and culture they worked in. Finally, among the coworkers included in the study, feeling a part of a community was related to thriving in their work environments. The researchers write, "Connections with others are a big reason why people pay to work in a communal space, as opposed to working from home for free or renting a nondescript office."

A study by the US Bureau of Labor Statistics found that contingent workers were twice as likely as noncontingent workers to be under the age of twenty-five. Thus, nonstandard work may very likely become the standard form of work for millennials and Generation Z. Young creatives pursuing diverse entrepreneurial paths often populated the coworking spaces that I observed. As more people join the nonstandard workforce, as freelancers, self-employed workers, contract employees, or gig workers, they, in all likelihood, will be working in a variety of nonstandard environments. Some of these environments will certainly be coworking spaces.

PLAYING TOGETHER, HUSTLING TOGETHER

During my research I worked closely with a sociology PhD student to conduct ethnographic research in two coworking spaces in Austin, Texas. I became especially intrigued by a coworking space called Chicon Collective. If we were going to create a coworking space that best embodies Jane Jacobs's edict that new ideas need aged buildings, we could not find a better example than Chicon. The two-toned aged brick building looks out of place in a neighborhood undergoing rapid gentrification. In a previous life the building was the rustic locale of East Austin's old Third Street Bus Depot. A couple of blocks south is an avenue bustling with new farm-to-table restaurants, yoga studios, and chic hangouts that offer craft beer to the influx of young college-educated whites who are settling in East Austin.

The members who gathered at Chicon referred to themselves as a motley crew of designers, makers, hackers, artists, musicians, programmers, and entrepreneurs who, in the words of their manager, were "working together to make the world a little more beautiful and a lot more interesting."

Despite some obvious renovations and repairs, Chicon's interior presented a hardscrabble look compared to some of the more upscale coworking spaces in town. But the building closely resembled the lives and careers of most of its inhabitants: incomplete and in constant iteration. Many of the members took great pride in the gritty aesthetic of the building and, more importantly, in the people who made Chicon the physical and social hub for the pursuit of their creative, entrepreneurial, and civic aspirations. The culture at Chicon reflected the tenacious ethos of the new innovation economy that many young creatives are drawn to.

Like most coworking spaces, Chicon offered its members access to a few private offices, open desk space, a conference room, speedy Wi-Fi, a recreation room, and a kitchen. The front part of the building could serve as a walk-in workspace by day and a small but functional event space by night. Most importantly, Chicon also offered its members access to other young creatives navigating an uncertain economy.

During our fieldwork, the group of workers at Chicon was made up mostly of young men with an inclination for programming, tech, and design. The graduate student who conducted most of our ethnographic observations of the space noted that Chicon had a "fraternity house feel." There were some women entrepreneurs in the space, but their male counterparts heavily outnumbered them. As a result, life in Chicon was symptomatic of the male hegemony that typically dominates the tech and entrepreneurial world.

What Chicon lacked in corporate-style amenities and well-connected players in Austin's entrepreneurial scene, it made up for it with a spirit of sharing and community that was central to its identity. One member said that he liked Chicon because "it's a community that gives a shit." He told us about the time when an aspiring entrepreneur came into the space and several of the members sat around a table and gave him feedback for two hours on his start-up.

"Even after the guy had left for the day, we still kept brainstorming ways to make his start-up better without any kind of monetary incentive," he noted.

This spirit of cooperation and support made Chicon different from some of the more competitive coworking spaces we encountered in our fieldwork.

"We've had people come in who are competitive," a member of Chicon told us, "but they usually don't last that long."

The coworking space was flexible, conducive to iteration, and open to a variety of activities. Depending on the hour of the day that we visited, it could be difficult to tell that Chicon was a coworking space. The members regularly put on community-building events like "Taco Tuesday" or bourbon tasting that kept them connected and in high spirits, literally and figuratively. One member offered an occasional yoga class, which was another clever way to form bonds while also promoting health and wellness. Chicon workshops taught everything from pickling to coding. In the late evenings Chicon could become a place to hang out and share takeout or drinks with colleagues and friends. The large projection-screen television on one of the main walls was used to play video games or watch sport events like the World Cup or NBA basketball. On designated nights or weekends Chicon might be the site for a meetup, a seminar for coders, a workshop for artists, or a boot camp for those practicing the side-hustle lifestyle.

In the afternoons Chicon resembled a coworking space—a mix of young twenty- and thirty-somethings banging away on their laptops, sketching ideas on whiteboards, conducting Skype calls, consuming massive amounts of caffeine and snacks, and looking exhausted from the unforgiving pace of hustling to create their own future.

A BEAUTIFUL STRUGGLE

We met a lot of interesting characters at Chicon, including a young African American social entrepreneur named Michael Henderson. Like his colleagues, Henderson labored to turn Chicon into his very own innovation lab—a place, that is, where he might defy the long odds of making it in an Austin innovation economy more hospitable to young creatives and entrepreneurs who fit a specific profile: white, male, college-educated, and intimately connected to the "bro culture" sensibilities dominant in the tech economy. Henderson used Chicon to grow and test his ideas, build his social network, hold events, and learn from other entrepreneurs while also enhancing his prospects for translating numerous hustles into a viable living.

Henderson was one of the few African Americans that we met in the Austin tech and start-up ecosystem. He was young, smart, wide-eyed, and ready to make his mark on the world. He loved the idea of pursuing his own vision of enterprise. Every day, he believed, was an opportunity to make his future. When we met Henderson, he was involved in a few different projects. He was just launching a new social enterprise to bring technological solutions to the developing world. He had also recently teamed up with an illustrator to design and produce a children's book. The idea for this project emerged after he was unable to find a children's book with a Black central character for his niece.

From what we were able to discern, hustling was not Henderson's side gig—it was his only gig. After graduating from Howard University, he worked on the first Obama presidential campaign in 2008. "I organized all of northeast Philadelphia," he said. "After that I worked on the inaugural committee, went to the ball, and wrote a proposal for the Smithsonian National Museum of African Art." The proposal was successful and led to a $3 million grant. Henderson thought the new grant would guarantee him a job, but in his words, "Political stuff started happening, and it just kind of got washed away." Faced with few employment prospects in DC, he moved back home to Austin. Moving back in with his mother made for a softer landing while he figured out his next steps.

A friend introduced Henderson to Chicon, and he liked the vibe and young creatives who worked and played there. It was at Chicon that he began hatching an idea to launch his own social innovation start-up called Developers Doing Development. The goal of the start-up was to design ways to use technology to solve problems in the developing world.

As I thought about young creatives like Henderson, their entrepreneurial pursuits made sense, even though the odds for success were long. Henderson was obviously ambitious and talented, but he did not pursue social innovation because he wanted to be rich or famous; he pursued it, in part, because there were so few meaningful employment opportunities in the labor market for him.

While Henderson never mentioned the lack of diversity in Austin's start-up and tech economy, his eagerness to launch his own enterprise was tempered by the reality that Austin's entrepreneurial ecosystem is marked by homophily. Like many tech hubs in the US, Austin's was overwhelmingly white and male, and like other innovation hubs, the more prosperous it grew, the more unequal it became. Young creatives like Henderson faced the added burden of how the culture of exclusion and lack of diversity limits opportunities for African Americans, Latinos, and women in the formal innovation economy.

One of Henderson's close friends and colleagues at Chicon was a young Nigerian named Moyo Oyalola, who moved to Austin with his family when he was a young child. Henderson and Oyalola were both in their late twenties and living at home with their parents. They were often unsure of where their next gig or source of income would come from. Oyalola was a little older than Henderson, though his short height, small frame, and boyish look made him appear younger.

Oyalola was also a chameleon. Trained as a graphic artist, one day he might be working a gig as a photographer, and the next day he could be helping a start-up design a new website or a new advertising campaign. He also built furniture for Chicon. But his entrepreneurial and start-up activities brought little if any income at all. When we asked him what was the hardest part about being self-employed, his reply was consistent with most young people in a predicament similar to his.

"Besides the financial part and waiting for money to come?" he said, laughing. "Right now, I mean, it's just always kind of worrying about financials and thinking about, like, you know, when's the next gig going to come. I mean, I'm doing the start-up thing, and I'm also, like, freelancing on the side, because we're not paying ourselves." He paused for a couple of seconds before referring to the fledgling entrepreneurial pursuits and side hustling as a lifestyle. "It's what we have chosen to do. It just seems right," Oyalola said. "I worked for a year right after I was done with school, but I was kind of wasting away. I was like, 'This is not what my purpose is in this life. It expands beyond this.'"

Like many other young self-employed persons, Oyalola liked what he called "the freedom and not being confined." The young graphic artist explained that humans have been conditioned to think that a nine-to-five Monday-through-Friday workweek is normal.

"I see opportunity every day, whether it's a Saturday or a Sunday," he explained. "Each day I wake up I feel like there's an opportunity here that exists. So, like, just having that freedom and knowing that anything can happen at any given moment is what I love, versus, like, okay, I'm going to work here, I'm going to wait for my promotion, or wait for this happen." He added, "There's a lot of question marks, but it keeps me going."

Oyalola called the life of young creatives "a beautiful struggle."

"EVERY DAY IS A HACKATHON"

Michael Henderson and the small team that he worked with at Chicon were always trying something new, fusing together the raw pieces of everyday life to make their vision of social entrepreneurship a little more real. The playful atmosphere at Chicon belied the intensity of the daily struggle most of the members waged to realize their entrepreneurial aspirations. As Moyo Oyalola said, "Every day as an aspiring entrepreneur is like a hackathon."

Michael Henderson's first opportunity to realize his social innovation start-up had connections to a program from the Obama White House called the Young African Leader Initiative (YALI). The president recognized that growing Africa's tech and entrepreneurial talent was a key part of any strategy to bring more innovation to the continent. In 2014 the White House invited more than two hundred young African leaders and entrepreneurs to the US. They came from all over the continent, including Ghana, Nigeria, Kenya, and South Africa. In addition to attending a summit in Washington, DC, the guests were organized into small groups and sent to visit innovation hubs across the US, where they met employees at start-up companies, tech innovators, and university leaders. Twenty-five of the young Africans spent six weeks in Austin.

Henderson learned from a friend of a friend that the young African entrepreneurs would be spending time at the University of Texas and visiting some of the start-up and tech companies in Austin. When the friend asked Henderson if he could think of people the young African leaders should meet, Henderson immediately thought of Chicon and the community of young creatives there.

After some contemplation, he began planning an event at Chicon that included dinner for the young leaders, small presentations, a meetup, and networking. It was an opportunity to introduce the visitors to their counterparts in Austin, build a viable social network, and explore future collaborations in the social innovation space.

At the resulting event, Henderson and the other members of Chicon learned that the businesses created by the young African entrepreneurs were designed to inspire, educate, and empower other young Africans. Many of the female innovators, for example, invested their time and energy in enterprises that would improve the health, education, and financial status of women. The Obama White House launched YALI precisely because the Africans present at the event Henderson organized embodied the continent's young talent and future promise as a site of investment and opportunity for future enterprises.

After their evening with the young African entrepreneurs, Henderson and his team decided that any social enterprise in Africa should figure out how to leverage the widespread diffusion of mobile phones across the continent. "We found out that mobile phones are pervasive throughout the continent of Africa, specifically feature or 'dumb phones,' Henderson recalled. "We decided that we should focus on that."

Designing a mobile-based intervention to be deployed in Africa required developing both a technical solution and an understanding of the cultural and behavioral nuances associated with the adoption of mobile phones in Africa. To facilitate the process, Henderson and his team at Chicon decided to host a hackathon, coordinated events that bring together coders, engineers, artists, educators, and anybody else to create technical solutions to a specified challenge in a finite period of time, usually two to three days.

Henderson leaned heavily on his previous conversations with the young African entrepreneurs to ensure that the ideas and technical solutions developed during the hackathon were not only technically relevant, but culturally relevant too. More specifically, the point was to design solutions for everyday Africans—not the tech savvy Americans and Africans who were participating in the hackathon. In the professional world of design, this approach to problem-solving is called "human-centered design." The idea is that any solution must be designed with the users and not the designers in mind. A central component of this approach is something designers call "developing empathy," which is the ability to see the world through the eyes of those you are designing a product or service for.

The weekend hackathon produced a number of prototypes. One of the apps addressed the Ebola crisis in Africa. One of Henderson's partners, Scott Akers, told me, "We didn't go into this when we first started planning to say this is the Ebola hackathon." But as the hackathon evolved, the idea of helping health organizations better understand the on-the-ground situation and people's behaviors gained traction. Participants in the hackathon built a ten-question survey to evaluate the risks of someone contracting Ebola, based on their activities and their symptoms. Akers added, "We take that information, build out a real-time heat map of responses coming from this survey, and provide that to NGOs, government organizations, and other boots on the ground so that they can better use their resources."

Akers described another health-based app developed during the hackathon: "We created an application that builds out infrastructure so that people can take pictures with their phone, send it to a network of doctors across the world who can help evaluate visible symptoms and propose potential courses of medical care." These prototypes explored the capabilities of mobile-based and artificial intelligence solutions in the developing world.

The small team of designers, developers, and social innovators developed these apps largely through the inventive use of a coworking space and their tech and social ingenuity. If it was a struggle to coordinate and actually pull off the hackathon, the transition to next steps—implementing their solutions—was equally daunting. Akers noted that the biggest challenge was "getting eyeballs on what we're working on." This was a challenge that resurfaced throughout my conversations with young creatives. Henderson and his team reached out to as many different stakeholders as they could through their social networks, social media, and the local press. They even received a favorable write-up in the *Huffington Post*.

Many of the young creatives who I discuss in this book—social innovators and media makers—do not lack ideas or ingenuity. Rather, they struggle to access the resources to realize their ambitions in a way that attracts financial investors, generates attention, and achieves sustainability. It is frequently said that in the anytime, anywhere media environment that we live in today, content providers are competing for their share of constantly moving eye-balls. Thus, in a world that is spawning more designers, media-makers, and entrepreneurs than ever before, standing out in today's attention economy is a daunting challenge.

A few weeks after the hackathon, I sat down with Henderson and his team at Chicon to get their assessment of the event. The forty-eight-hour hackathon produced four prototype mobile solutions to address some of the critical needs of communities across Africa. I asked Moyo Oyalola what, in his mind, made this hackathon different from others he had attended. "Obviously the focus," he replied. "Not a lot of people traditionally do hackathons for feature phones, plus contextualize it for the African continent."

He continued, "And I think things like that open it up to bigger dialogue about what is the African content. What's going on there? And that's really how you kind of start changing the narrative of what's possible." In order to change the narrative about Africa, Henderson and his team needed good ideas, which they had. In order to test, iterate, and bring their ideas to market, they also needed financial resources, which they did not have. As a result, their desire to use mobile phones to enhance life on the African continent stalled.

A few weeks after talking with me, Henderson and Oyalola traveled to two African countries to explore ways to collaborate further with their African counterparts. Henderson even took a copy of the children's book he had created, to share with children at an African school he planned to visit. Their meetings at Chicon with a group of young African entrepreneurs had introduced them to a new world of possibilities, one that took them a long way from Austin and the coworking space that housed the Chicon Collective.

YOUNG CREATIVES AND THE MEANING OF MEETUPS

The North Door is one of the many places that we visited during our field-work that embodies the distinct geography of the new innovation economy. Several decades ago the stone building that is home to the North Door was originally a buggy factory and then a feed store. In the 1970s an aspiring filmmaker named Richard Kooris and his wife, Laura Kooris, moved into the building to start a film production facility. At the time, the idea of a film production company in Austin seemed, well, just plain weird.

According to Laura, they "resurrected [the stone building] from despair into a creative think-tank and facility." As their little film production company began to grow, it became a hub for filmmakers and their ancillary partners, such as editors and sound designers. Over the years, the building became a soundstage for the filmmakers who built an independent film industry in Austin literally from the ground up. Years later, other creative types, such as music producers, graphic artists, and nonprofits, began working out of the space. To accommodate the growth the Koorises purchased three additional buildings that stretched across two blocks. Today the complex is called 501 Studios, and it is a throbbing center of creative activity and entrepreneurship.

Today the North Door brands itself as a unique event space that is ideally suited for a generation of creatives looking for a place to get their hustle on. On any given night you might see a local indie band, an up-and-coming national musician, or spoken-word artists performing onstage. The twenty-five-foot HD video projection wall makes it an ideal space to screen indie films. The two bars and balcony seating add to the spirited vibe that makes the North Door the perfect event venue for aspiring creatives and those who take pride in patronizing the new and the next.

Every first Thursday the North Door plays host to a monthly meetup of indie game developers organized by the founders of a collective called Juegos Rancheros. Juegos formed as a resilient response to the game industry layoffs that left a number of young creatives in Austin searching for their next opportunity to make interactive media. Game development in the Juegos collective was everybody's side hustle, which meant that they were usually making games late at night, on weekends, and whenever else they could find time. The collective was an attempt to establish a mechanism for young creatives to connect with one another, grow their social ties, and strengthen their capacity to make games.

Wiley Wiggins, an independent game developer and Juegos cofounder, told us, "You need people to talk to and people to ask for help. And that was always kind of the idea . . . that we would be introducing talented people to one another, maybe even people outside of games—musicians, artists, and animators."

The cornerstone of the Juegos creative ecosystem was the monthly meetup at the North Door that brought all kinds of creative types together, including game developers, programmers, artists, writers, musicians, and filmmakers. Meetups, I learned over the course of my research, are a key feature of the new innovation economy that young creatives are building. Meetups like the one organized by Juegos are another example of how young creatives are dispelling the myth that their attachment to technology makes them antisocial and disinterested in their peers and community. Like most meetups that we observed, Juegos uses them to help participants grow their social connections in ways that, among other things, stimulates their creativity and expands their opportunities to pursue their game-making aspirations.

The meetup idea is certainly not new, but it has taken on a distinct resonance among young creatives striving to turn a side gig into a main gig. As more individuals work as self-employed workers, independent contractors, or freelancers, the likelihood of social and professional isolation increases. We visited the Juegos meetups on numerous occasions and noticed that the event stages regular opportunities for young creatives to connect, collaborate, and cultivate their identities and abilities as indie game developers. The new innovation economy, it turns out, has a social life.

WHY YOUNG CREATIVES EMBRACE OPEN INNOVATION

Several of the Juegos meetup participants were former game industry employees. While they liked the idea of working in the industry, we frequently heard complaints about how the dominance of commercial interests over artistic interests, nondisclosure agreements, and noncompete clauses prohibited them from talking about the games that they were working on or from developing their own games. The meetups established the environment for the participants in Juegos to practice a more collaborative form of innovation, what some researchers refer to as "open innovation."

There are many ways to think about the concept of open innovation, but at its core it points to how organizations, large and small, look to leverage the flow of ideas, expertise, and insights that circulate beyond their boundaries. Unlike the silo mentality of the well-funded corporate triple-A game studios, for example, Juegos' use of an open innovation model suggests a willingness to participate in a wider community of knowledge and expertise to drive game development.

During the Juegos meetups, for instance, indie game developers frequently shared the games they were building without fear of reprisal or violating a signed agreement. The meetups were often used to help brainstorm ideas for current or future projects. In short, most of the indie developers understood the benefits of sharing their games and ideas with their peers. They understood that making games in a closed environment limited their access to ideas, talent, and knowledge about game creation.

The open innovation model illustrates how the creative process is especially porous, happening between the different but interrelated realms of knowledge, expertise, and social interaction. Juegos' culture of open innovation was notably different from the more conventional and closed approach adopted by the triple-A game studios. Take, for instance, how the participants in Juegos often tested their games. Whereas corporate game studios test their games in more controlled and secretive settings, many Juegos developers opt to test their games in the relatively open environment of the meetup. In instances like these, the meetups function much like a pop-up testing lab to elicit real-time feedback from their peers about everything from game-play mechanics to artwork. The feedback facilitates iteration and game creation.

One young game designer described these informal play-tests this way: "Somebody goes in the corner, and they set up a laptop, and it's, like, they've been making this game for the last six months and they haven't shown it to anybody." He continued, "They feel safe and good, and they feel confident to show it to all their peers. And that is just, like, gold compared to an industry that usually is all about secrets, everything's proprietary. But here we want to share it with everybody, and we want our ideas to grow based on the feedback we get. And, you know, that to me is what making games should be about rather than just kind of, like, cutthroat competition."

The open innovation concept is spreading across the broader innovation economy, in the form of more open and collaborative work environments. It is also embodied in the growing popularity of events that bring people together to design solutions in the spirit of shared purpose and collective intelligence. A report by the Brookings Institution contends that established companies are adopting elements of open innovation by, for example, relocating to areas that allow their workers to cluster among other organizations, exchange ideas, and, thus, accrue the benefits of knowledge spillover. The open innovation concept suggests that when people from different disciplines engage in conversation with one another, ideas can flow and collide in ways that lead to new and promising insights.

THE REAL SHARING ECONOMY

The North Door meetup was also the site for Juegos members to participate in what I call the "real sharing economy." Global tech platforms like Uber, Lyft, and Airbnb dominate our view of the sharing economy. These companies represent what some scholars call "crowd-based capitalism." One of the core features of the Uber-style sharing economy is the supply of capital and labor from decentralized crowds of individuals instead of corporate entities.

This particular expression of the sharing economy is, however, market-based; it is predicated on the commercial exchange of services. As some critics have noted, the service-for-money transactions in this context do not really constitute sharing in the traditional sense of the word.

The sharing economy that young creatives are building is markedly different. It is an economy based on reciprocity and support rather than financial transactions and profit. In this sharing economy people literally share something—such as a skill, their time, or expertise—with someone else without any expectation of financial remuneration. I witnessed and heard about this form of sharing throughout my research. The practice of sharing is a critical feature of the new innovation economy. It is a smart and necessary use of one of the few forms of capital accessible to a generation of young creatives who are long on talent, ingenuity, and hustle but short on cash, staff, and organizational muscle.

Importantly, the rise of this sharing economy disputes the narrative that millennials do not invest in interpersonal relationships and their communities. Further, it contradicts the view that they live "alone together." In fact, this sharing economy only works when young creatives are part of and consistently contribute to a community. Let me offer one example of how the North Door meetup energizes the sharing economy.

When the founders of an independent game studio set out to launch a Kickstarter project, they tapped their Juegos-related social network to recruit the talent they needed to execute an effective crowdfunding campaign. As I discuss in greater detail in chapter 2, crowdfunding has become a competitive endeavor. It is not enough anymore to simply ask people for money. You have to present your campaign in compelling fashion, which includes telling a carefully crafted story, producing professional-quality video and graphics, and offering rewards that incentivize giving. Creating a crowdfunding campaign these days requires a level of talent and expertise that is often beyond the scope of what most budding entrepreneurs possess.

As the two cofounders began to develop their crowdfunding project, a Juegos colleague introduced them to an independent filmmaker who agreed to produce a video for their Kickstarter campaign. Also, through their connections at the Juegos meetup, they found an artist who produced a poster art series, prints of which the cofounders offered as incentives for individuals to give to their Kickstarter. Ordinarily, talent like this would cost money. But the social capital these game developers had built through the Juegos meetups facilitated access to talent and services that were basically free. This was not an isolated example of sharing. It was quite common at Juegos specifically, and within the new innovation economy more generally.

"It's not that we're just super charming and that everyone wants to help us; it's that we're good citizens in a community," one of the indie developers told me during an afternoon interview. "Like, when people need help with their stuff, we're happy to lend a hand or take a look and give an opinion, or just doing whatever." Her partner added, "I don't think our game would've come out anywhere near as good as it did had we not had all those folks."

The idea of helping and supporting by sharing time, talent, and ideas was a norm that had been established in this indie game collective. Like others in this community, these two game developers turned to the Juegos meetups regularly to elicit all kinds of support throughout the development of their first independently produced game. They made the game with virtually no money, no staff, and no prior experience developing a market-ready product. Without this informal economy of sharing, the development of their game would have been considerably more difficult.

For many of Juegos Rancheros' indie game developers, the sharing economy was a crucial resource that generated substantive benefits. More precisely, for the two developers launching the Kickstarter campaign, the sharing economy made an otherwise impossible task—building an indie game with virtually no money—possible.

The North Door was a crucial part of the side-hustle economy that these indie game developers participated in. The meetups were a creative use of a building that decades ago had been used to make buggies. Among the many things that this space made possible, none was more important than the social connections and community resources it afforded this group of hungry young creatives.

CHAPTER 2

BOOTSTRAPPING

Inside the Quirky World of Indie Game Developers

Jason Rosenstock, a graphic artist, sat down for an interview on a hot Texas summer evening. In his mid-twenties, Rosenstock had the look of someone eager to make it. His brown beard was full and well groomed. This evening he was dressed casually in a T-shirt and blue shorts. Jason had moved to Austin from New York to work as an artist for the triple-A game studio BioWare. The studio, owned by Electronic Arts, was among the many that had set up shop in Austin, eager to take advantage of the creative-is-cool lifestyle, the burgeoning tech economy, and the growing flock of highly educated young creatives who were making Austin their home.

"I had never worked in the games industry before, so I was thrilled for this opportunity," Rosenstock said. "I was really excited to get a job offer. It was like a dream come true. So moving down here for that job was a big deal."

Rosenstock was part of a migration wave that was remaking Austin's demographic profile. The city has long maintained a reputation as an innovative hub—places described by Enrico Moretti, an economist at the University of California at Berkeley, as cities with high numbers of college-educated workers and innovative employers. Austin is one of the cities Richard Florida ordained as a magnet for the "creative class" in his book *The Rise of the Creative Class*. In that book, Florida celebrated metropolitan areas that emerged as the landing spots for a new class of workers who signaled the shift from a goods-producing economy to a knowledge-producing economy.

But innovation hubs also come with noteworthy costs. Despite the economic benefits, these hubs have been heavily criticized for accelerating racial, social, and economic inequality. Even as Austin was adding educated millennials to its population, it was losing Blacks, Latinos, artists, immigrants, and young families with children who could no longer afford to live in the city. This is a reality faced by many innovation hubs, including San Francisco, Seattle, New York, and Boulder, just to name a few.

KNOWLEDGE INDUSTRY SWEATSHOPS

Buoyed by a sense of creative purpose and optimism, a number of game studios set up offices in Austin and went on a hiring spree in the 2000s. This included major media corporations like Disney, Electronic Arts, and Blizzard Entertainment, as well as upstarts like Zynga, known for its achievements in social and mobile gaming with titles like *Farmville* and *Words With Friends 2*. Between 2005 and 2009 the Texas game industry grew by an annual rate of 13.7 percent, roughly five times the rate of the state's overall economy. By the early 2010s, only one state, California, employed more people in the game industry than Texas.

BioWare opened up an Austin office to build *Star Wars: The Old Republic*, a massive multiplayer online game. In 2007 Disney acquired the Austin-based Junction Point Studios to establish a presence in the world of gaming with the production of *Epic Mickey*. Blizzard Entertainment established a footprint in Austin, as did Zynga. But just when the Austin gaming economy appeared to be on solid ground, the bottom fell out.

After releasing *Star Wars: The Old Republic* in 2011, BioWare began laying off a significant number of the people it had hired to produce the game. When Zynga announced that it was laying off about 5 percent of its employees in 2012, the shop in Austin was hit especially hard. In 2013, after disappointing sales of *Epic Mickey*, Disney announced that it was shutting down Junction Point Studios. The effects rippled throughout the Austin game industry. "Suddenly, many of my peers were being told we didn't have jobs anymore," recalled Rosenstock, who was no longer retained after his contract.

Employment within the knowledge economy is seldom glamorous. As tech companies have grown in terms of financial earnings, stock valuations, and cultural influence, they have also come under greater scrutiny for their toxic organizational behavior and lack of racial, ethnic, and gender diversity. A *New York Times* exposé in 2015, for example, revealed the toxic workplace culture at Amazon. Several reports have exposed the "bro culture" in tech that makes it hard for women to earn respect, promotion, and salaries comparable to their male counterparts. The game industry mirrors many of the workplace culture and diversity issues that typify the broader tech sector.

The 2017 *Developer Satisfaction Survey* produced by the International Game Developer Association (IGDA) found that 74 percent of game developers are male. Further, 68 and 18 percent, respectively, are white and East Asian. By contrast, Latinos make up 5 percent of developers, and African Americans make up 1 percent. The IGDA concludes that the prototypical game-industry worker is a thirty-two-year-old white male with a university degree who lives in North America.

Moreover, the game industry has a notorious reputation for overworking its employees. Countless stories have been told about the industry requiring employees to work excessive numbers of hours per week to deliver a game title on schedule. People in the industry call this "crunch." When asked about workplace conditions in the industry, game developers confirm the reality of crunch in the 2017 IGDA game-developer survey. More than half, 51 percent, indicated that their job involves crunch time. The number of weekly hours worked during crunch varied from between fifty and fifty-nine hours to in excess of seventy hours. Even as games are characterized as a prestige industry, the long and sometimes grueling hours in the sector can resemble the conditions in a sweatshop.

Knowledge industries like game development are also famously fickle. Jobs in the industry come and go. The massive studio layoffs in Austin that began in 2011 were not isolated incidents or events spurred by recession. Besides, practically all of those working for the studios in town were working as independent contractors with contracts that specified the length of their employment. While some observers celebrate the "flexibility" that contract employees have in the so-called gig economy, the reality is that employment arrangements like these usually favor the employer. The 2017 IGDA report found that the average game-industry employee changes jobs every 2.4 years. It also explained that permanent employees "are often hired and let go." Their job security, it turns out, is not much better than those who work in the industry as freelancers.

While the game industry is touted as an exciting place to work in a bourgeoning tech and media industry landscape, the industry is actually a classic example of what Matthew B. Crawford calls "cognitive stratification." According to Crawford, enthusiasts tend to overlook the fact that most people employed in tech and media are not working in jobs that allow them to express artistic, creative, design, or leadership skills. For example, in the game industry, the bulk of the people are employed in service-oriented occupations that support the few who actually work in the more prestigious positions: those who manage workers, design the games, create the art, and market the multimillion-dollar titles produced by the big studios. The game industry illustrates the stratification that takes place among knowledge workers in other industries.

Trends like contract-based employment arrangements, crunch, and cognitive stratification underscore the degree to which work in the new knowledge economy can be just as soul crushing as work in the old industrial economy. What's even more frustrating for many young people entering the knowledge economy is that they are doing so with college degrees that they'd hoped would lead to fulfilling work or upward mobility. It is not simply a case of paying your dues as you work your way up the company ladder. In more and more instances, there simply is no ladder to climb. For example, the contract jobs that many young creatives took in Austin's game economy were the norm and rarely provided a pathway to occupational mobility or higher pay within the company.

Many of the former game-industry employees that we met still felt the sting from working in the industry. Many had moved to Austin to pursue their dream job: making games. Despite the layoffs and the subsequent reshaping of the industry's footprint in Austin, many elected to stay. A few of the displaced workers landed jobs in other creative and technology sectors, but most struggled to find employment that was commensurate with their skills, education, and expectations.

A HUB FOR INDIE GAME DEVELOPERS

The young creatives displaced from Austin's games sector responded creatively to their predicament. One example is the creation of the independent game collective Juegos Rancheros, discussed earlier. The collective was founded by Brandon Boyer, a game journalist and the chairman of the Independent Game Conference; Adam Saltsman, a game developer and CEO of an indie game studio; and Wiley Wiggins, a user-experience designer and independent game developer.

The founders of Juegos realized a need for some kind of collective activity in the wake of the widespread studio shutdowns that pushed a number of young creatives out of the industry. Faced with a reduced games sector footprint in Austin and an uneven demand for tech workers, they sensed an opportunity to build an independent game-development ecosystem. Before

Juegos, there was no organized mechanism for the displaced workers and the growing community of young creatives to meet regularly, network, share ideas, and catalyze indie game development. Juegos' anchor event—the monthly meetup discussed in the previous chapter—was a creative solution to this dilemma.

The objective of Juegos was both modest and ambitious: to create a hub for developers, artists, and other creative types to come together and build an indie game development community that, among other things, fostered opportunities to network, collaborate, learn, and make games. Like so many ideas that have sprung up within the new innovation economy, the idea that became Juegos was both inventive and iterative from the start. In just a few years, Juegos evolved beyond the monthly meetup to host local and global game events. Juegos also became an important national hub for indie game development and the organizer of an indie game developer festival that has grown its profile by attracting indie game developers from all over the world, as well as blue-chip sponsors like Sony. The members of the Juegos collective exercised tech and social ingenuity to build a distinct economy of innovation.

BEING INDIE

We asked several of the indie game makers affiliated with Juegos to define what it means to be an indie game developer. They offered many definitions, which suggests that indie game development is a diverse enterprise and one that defies easy descriptions and categorizations. Several of the indie game developers that we spoke to likened indie game-making to the practice of indie filmmaking. Film was once the lone domain of deep-pocketed studios. But advances in technology and distribution have lowered the barriers to entry for making and screening films. For example, affordable high-end cameras and editing tools make it easier for indie filmmakers to produce professional quality films. The emergence of alternative distribution channels, such as film festivals and the internet, has strengthened the ability of indie filmmakers to circulate their work. A generation of filmmakers who believe they have something to say through the medium also drives change in the industry. These and other factors make it possible for aspiring filmmakers to produce films that are personal and professional while also challenging industry conventions in terms of style, narrative, and representation.

The ability to make high-quality indie video games that command even a small degree of attention in a global gaming economy is a recent phenomenon. Historically, the triple-A game studios have benefited from what has been a virtual monopoly on the resources—money, technology, and talent—to make, market, and distribute games. The big studios continue to dominate the market, but notable technological and social shifts have forged an alternative economy for indie game developers to exploit.

First, the emergence of relatively affordable software like Unity, Maya, and Unreal allows indie developers to make games with stunning graphics and challenging game play mechanics. Second, social platforms like YouTube and Facebook allow game creators to develop trailers and other media that promote their games while also building a community. Third, distribution channels like Steam and itch.io, and the mobile app economy make it possible for indie game developers to share their games in the long-tail economy made possible by the internet. The Long Tail theory suggests that as the internet makes distribution and consumption easier, demand will shift from the most popular products to a host of niche products along the tail end of the demand curve. Finally, a new generation of game-makers, represented by the members of Juegos Rancheros, is asserting its own notion of games.

Many Juegos members created studios to support their desire to make games. Independent game studios come in many different forms and sizes, including big, medium, and small. Jo Lammert, an indie game developer, explained that "indie" was a flexible term that did not fully capture the differences among the indie game studios in town. "Like here in Austin, you have indie studios that are literally one person. And then you have teams like Certain Affinity, which is totally technically indie," Lammert told me. "They work on things like Halo and other giant intellectual properties, but they're their own, you know, self-contained studio." She noted that Certain Affinity was not a publisher but that "they just work on some different parts of other projects, and then sometimes work on their own projects. But they're totally indie."

Speaking about the term "indie," George Royer adds, "I think it sort of means what 'alternative' did to music around 1999. It just means you don't have a publisher."

The indie studios that we encountered were all small shops, usually consisting of no more than two or three people. Their creators called them micro-studios. None of the micro-studios had a budget of any significance to make games. The developers typically made games with teams small enough to fit into a small kitchen. One word came up consistently in our conversations with indie game developers to describe their studios and their games: "bootstrapping." The word aptly describes an ecosystem populated by game developers who are ambitious but always climbing uphill. Bootstrapping acknowledges that indie game making is a scrappy enterprise made up mainly of hustle, persistence, and ingenuity. Indie game development among the members of the Juegos collective was generally a side hustle, something that they did alongside a more permanent job that paid their rent and put food on the table.

The bootstrapping ethos that was core to indie game development in the Juegos collective was a sharp contrast to the culture in triple-A studios. Whereas the corporate studios are defined by multimillion-dollar production and marketing budgets, the micro studios we encountered make and market their games with virtually no money. Triple-A studios are defined by large teams made up of artists, programmers, graphic designers, sound and audio experts, special effects technicians, and marketing personnel, just to name a few. It was not uncommon for a studio in the Juegos ecosystem to consist of one to three people who usually possessed some skills, but nowhere near the scope found in the big studios. Finally, unlike the big studios, the indie game-makers had no connections to publishers, hence they sold their games via the internet and as apps in the long-tail economy made possible by digital distribution.

These obvious disparities notwithstanding, indie game developers told us there are some notable advantages associated with their marginal position. Jason Rosenstock, the graphic artist, explained one advantage this way: "For me personally, the main thing is that—and this is sort of a technical thing—you are not tied down to things like noncompete clauses, and you are able to freely work and collaborate with your peers and other people in the community without having to worry about such clauses."

Randy Smith, founder of a micro-studio, also believed that the small size and scale of shops like his offered important advantages. "The bigger games have to make a lot of money back, so they have to appeal to everybody," Smith said. "And they usually have to be sequels, and they're kind of stuck doing some of the same things that you see over and over in video games." He liked several of the games made by the triple-A studios but believed the pressure to make massively popular games stifled creativity. He also observed that triple-A studios overlook the fact that the average age of gamers, thirty-two, means that tastes in games likely evolve. "After you get used to the same old first-person shooter mechanics for the zillionth time, tastes mature a bit," Smith said. Triple-A studios are not nimble enough to account for the fact that as gamers age, they may develop more nuanced tastes that are seldom reflected in triple-A games. This gap leaves plenty of market space for indie game developers.

One of the advantages of making games on the margins of the commercial industry, Smith noted, "is that you don't have to make so much money back, because you're not spending hundreds of millions of dollars, and so you get to be much more personal, much more unique, much more innovative and artistic with your goals and your aspirations." This fact allowed him and many of his colleagues in this community to make, in his words, "games that no one's ever seen before—games that are more niche that'll appeal very much to some people but not to everybody." Smith described this position as a "nice place to be, because you get to be a little bit more authentic and have more genuine expressions in your work."

GAMECHANGER: THE UNITY ENGINE

The micro-studios that grew out of the Juegos ecosystem adopted the do-it-yourself ethos of the collective to make their games. The DIY approach to media-making has a much longer history in other industries, including music, video, film, and photography. The emergence of cheap tools in digital arts, design, and, interactive media has opened new frontiers in which indie game developers thrive. Photoshop and a free comparable alternative called Gimp are useful for images. A program called ZBrush is popular for 3-D sculpting. For animation there is Blender, which is also free. And for sound design many in this ecosystem use open access or cheap tools like Freesound.org, Apple's GarageBand, or Audacity, a free and open-source digital audio editor and recording application software.

YouTube is also an important platform for indie game makers. In its rise as the world's biggest video-sharing platform, YouTube has become a significant player in gaming culture. In 2017 half of the eighteen top-subscribed-to YouTube channels in the world were gaming related, according to *Business Insider*. For the indie game developers, YouTube is the perfect vehicle to showcase glossy trailers that help promote their games, build their brands, and foster a global community of indie game consumers.

But of all the tech tools that have buoyed indie game development, the real game-changer is Unity, the cross-platform game engine that was officially released in July 2005. The makers of Unity anticipated the disruption that was coming to the gaming world. More specifically, they foresaw the arrival of casual gaming across multiple platforms and, just as important, the emergence of a new generation of game developers who were not tied to the corporate studios. Unity, more than any other technology, has made it possible for independent game makers to develop high-quality 3-D games. An indie game developer can make a game and, using cross-platform functionality, release it for the web, iPhone, iPad, Android, Wii, PC, Xbox, and PlayStation.

From the start, Unity was a hit among aspirational game makers. Nearly every member of Juegos that we talked to considered it the standard in indie game development. They liked the quality of the technical and aesthetic features in the games they could produce. Unity was the tool that many of them had been looking for. Today, more games are made with Unity than with any other platform in the world. More than 700 million people around the world have played a Unity-built game.

The Unity software itself offers an elegant suite of tools and a workflow that are easily accessible to game developers. There is a free version and a more advanced version behind a paywall, but the free version provides a surplus of tools. For advanced developers, a professional version offers a free trial before payment is required.

This wide inventory of software gives indie developers powerful studio-like capacity to make high-quality games using their laptops in informal workspaces such as a place of residence, coffee shop, or coworking space. Today, there are more indie game developers in the world than developers working in triple-A studios. Consequently, the future of games is just as likely to be shaped by indie developers using free, open-source, and affordable software as it is by developers laboring in corporate game studios using more expensive proprietary applications.

EXPANDING OUR NOTION OF GAMES

The games made by members of the Juegos collective were purposefully distinct, even among the eclectic universe of indie game developers. Most of the members viewed indie game creation as an opportunity to express very personal and creative ideas about what games could be. More specifically, they viewed game creation as an opportunity to rethink the medium. This often meant thinking creatively and critically about aesthetics, story, game play mechanics, and even the politics of games. For instance, some of the developers used indie game creation to rethink two of the most dominant tropes in games: sexism and violence.

Rachel Weil was fully aware of the gendered history of games and its implications for girls and women. The young creative grew up playing games and was always uncomfortable with the portrayal of women in that context. "Right around junior high, that was when I got an internet connection and started researching games and emulation and things like that," she told a member of our research team. "And I got interested in game hacking and NES [Nintendo] game hacking in particular. And once I started learning how to hack games, I became really interested in the idea of making my own NES games."

Weil approached game design as a way to interrogate the sexism in games. She began visiting different online forums and ordered books about 6502 assembly language to deepen her knowledge about computer programming. The 6502 assembly language allowed individuals to program for the eight-bit microprocessor used, in this case, in early home gaming systems like Atari and Nintendo. "I taught myself how to code for the NES," she said. "In creating my own games I was able to sort of change the narrative about what games were about."

One of the first games that Weil decided to hack was Super Mario Brothers for the Nintendo NES. "I put Hello Kitty in, and I made a very cute version of Super Mario Brothers, because I was interested in sort of changing this narrative and recreating an NES or '80s or '90s gaming culture that was very welcoming to girls." From her perspective, a series of games in the 1990s that was more hospitable to females could have influenced a generation of girls to not only play games but make games too. Now Weil wanted to reimagine gaming culture's past in order to remake its future. "With hacking tools and a new sort of home brew and programmer cultures, we can reshape those narratives or change them in ways that reflect something we might've liked to have seen in the '80s or '90s," she said with an inflection of hope.

Weil notes that shifts in gaming platforms have ushered in a new era in gaming that makes it difficult to ignore the presence of girls and women as gamers. According to the Entertainment Software Association, girls and women make up 47 percent of all gamers. It is this rise in the percentage of female gamers that motivates developers like Weil. "You know," Weil pondered out loud, "so often we're excited about new technologies and what's the aesthetic of a game and how do you program it and all of these important sort of logistical points. But I think it's also important to think about the sort of social impact that a game might have. Why do you want to make a game? What sort of message do you want to convey with this game?" For Weil, the new innovation economy sparked by indie game development is not only an opportunity to make games; it is also an opportunity to hack the male gaze that has dominated game development and gaming culture.

The dominance of the first-person shooting genre in gaming is something that many Juegos members want to disrupt as well. "Everything that we try and do at Juegos is about breaking out of this gamer monoculture of a bunch of white dudes hunched over an Xbox shooting each other and calling each other names," game designer Randy Smith said. His studio is interested in making games that detour from the staple genres in which the triple-A studios invest hundreds of million dollars. For example, his team made a game called *Waking Mars*. Like many of the other games made by the Juegos community of developers, *Waking Mars* is a deliberate rejection of the violent action genre that dominates the roster of games produced by triple-A studios.

"This game is kind of a hard sci-fi game set on Mars in the real world in the future kind of a thing. In the game you never kill any aliens, you don't fight with laser guns, there's no acid-dripping fangs," Smith explained. "You do the exact opposite—you have to create life. So as you go through these frozen ancient caves, you waken the ancient ecosystem back to life, and then you slowly repopulate. You grow these ecosystems that interact with creatures and soil, and you learn how to create new life forms." The idea that a video game hinges on the idea of creating rather than destroying life is an example of the out-of-the-box thinking that goes into the development of the games made in this small but ambitious community of game makers.

WINNING THE SIDE HUSTLE

Members of the Juegos ecosystem began to pursue the development of indie games just as the industry was undergoing a radical transformation led by new software, the arrival of the smartphone, and alternative distribution platforms. Combined, these factors led to significant disruptions in the games industry and generated new opportunities for indie game developers.

Adam Saltsman is an indie game developer who maneuvered to exploit this shifting landscape. A cofounder of Juegos Rancheros and a central node in the ecosystem it cultivated, Saltsman is a celebrity among many indie gamers. He had managed to turn his side hustle—making indie games in a small but enterprising studio out of his home—into a full-time gig.

A native of Michigan and a University of Michigan graduate, Saltsman moved to Austin shortly after finishing college. The pull of Austin's burgeoning tech and entrepreneurial economy convinced him that his prospects for employment would be better there than in other cities. He found a software job in his newly adopted home. Even though it was not the job he coveted coming out of college, the software gig offered one crucial perk: "It was only forty hours a week," Saltsman said. "The real game industry jobs still had quite a reputation for sort of taking over your whole life," he added, with an oblique reference to crunch. "You know, essentially being forced to work sixty or seventy hours a week for months at a time in order to get the game done. And it's a really counterproductive practice, but it's still widely accepted in the industry."

Saltsman coveted working at a studio, but he realized that doing so would prohibit him from making his own games. The nine-to-five hours at his regular software job gave him time to freelance and pursue side projects. So he did. As a freelancer, Saltsman made 3-D models for games and pixel art for mobile phone games, and he worked on graphic design for many different projects. This side work allowed him to begin building a portfolio that showcased his skills for potential clients. Equally important was the fact that the freelance gigs helped him build the kind of social and professional network that is critical for gig workers.

Eventually he landed a few freelance projects that involved making 3-D models for clients, but in his words, "It still was not game design." He concluded that he had to shift from building a portfolio of digital artwork to a portfolio of digital games. "I started building little bits of Flash code that would let me put pixel art into the games instead of having to make really detailed drawings, and, you know, put simple sound effects into the games, and automate some of the simple animation processes that I use in a lot of games."

After about three years of working at his software job and maintaining his side hustle, Saltsman decided to quit his day job and began freelancing full-time. "It was pretty rough. I was doing a lot of contract work. I was working on something like twenty or thirty small games a year. You know, mobile phone games, and coming in and fixing art for old games, or prototyping stuff for EA [Electronic Arts], or, lots of weird combinations of things. I would program a server and Java interface for Satellite Radio one time, and then a thing that let you do magic deck building for a Facebook game."

Saltsman's nonstop side hustling eventually led to what freelancers call "turnkey" projects, or when one person or studio works on a whole job. Saltsman describes this transition as "the pivot," the moment when his small enterprise began to take on jobs that required making complete games rather than just the art, sound, or some other component. Turnkey projects were a win-win. The jobs paid more while also burnishing Saltsman's credentials as someone who could build entire games.

One of the simple games that Saltsman made on the side significantly elevated his profile in the indie game developer community. In 2009 he participated in a game jam sponsored by a group of Carnegie Mellon students called the Experimental Gameplay Project. Games jams bring people with different expertise together to find a solution to a specific problem in a finite amount of time. The core idea is to connect participants with different skill sets—animators, programmers, writers—to build games quickly and in a spirit of collaboration and experimentation. A game jam may be organized to develop games that address a specific challenge, for example in the civic or education space. In other instances, a game jam may experiment with a new game play mechanic, design technique, or piece of software. The main point is to generate an idea for a game and build it fast. But it is also a good opportunity to perfect skills and try out ideas you ordinarily might not.

The organizers of the Experimental Gameplay Project have some very simple rules: Each game must be made in less than seven days. Also, a single person must make each game. The main goal is to discover and rapidly prototype as many new forms of gameplay as possible. Saltsman benefited from the "make fast" ethos typical of game jams. He was forced to build the basic architecture for his game in five days. The constraints encouraged a simple design and aesthetic. As he proceeded with refining his prototype, the mix of design simplicity and game play intensity was very intentional.

"I thought, 'Let's make simple art that's really iconic, make really, really simple controls. But then the gameplay itself, it's, like, make the top of the game really simple and casual, but then everything underneath can be hairy and difficult and challenging and exciting,'" he explained. The result of Saltsman's work from the Experimental Gameplay Project was his prototype for the indie game *Canabalt*, now a cult classic. His minimalist approach to design was not the only thing that stood out about *Canabalt*. In a flattering piece in the *New Yorker* that highlighted the savvy design, game play mechanics, and cultural impact of *Canabalt*, writer Simon Parkin notes that Saltsman effectively reconciled "the smartphone's absence of buttons with the interactive complexities of contemporary video games." Parkin notes that "*Canabalt*'s solution was elegant and simple: tap anywhere on the glass and your character leaps." *Canabalt* was credited with helping to inspire a new genre of smartphone games, what is now called "the endless runner."

As a small indie game developer, Saltsman had no money to conduct a publicity campaign for *Canabalt*. So he did what most young creatives do: he mined the internet's crowd-building capacity to grow a community for the game. He posted a brief message about *Canabalt* on an online forum for game developers and made the game available on his website. Soon he received the most effective form of advertising—word-of-mouth—from indie game developers and consumers all around the world. The game went viral and was made available via several digital and gaming platforms. *Canabalt* strengthened Saltsman's street cred in the world of indie game development. The success of the game brought what he described to us as "middling celebrity." Success bred success, and soon he had access to a more dynamic social network and a range of indie game projects to develop. As a result of the success of *Canabalt*, Saltsman was doing something that is rare among indie game developers: he was winning the side hustle.

THE OTHER SIDE OF THE SIDE HUSTLE

In popular culture the side hustle is often glamorized, and its side effects are typically airbrushed out of view. In one of Uber's first television spots, a young driver tells viewers that Uber makes it easy to "go from earning, to working, to chilling, with the push of a button." In the spot, we see him driving a passenger, working at his day gig, and also enjoying plenty of leisure time in the pool, playing pickup basketball, and smashing a piñata at a party. Using the tagline "Get Your Side Hustle On," Uber makes the side hustle look fun, convenient, and hip. Popular television shows like *Two Broke Girls* and *Girls* seamlessly integrate side hustle story lines into their plots, minus the mental and physical toll. Hip-hop culture has built its whole identity and cultural economy around the side-hustle ethos. Rappers as varied as Jay-Z and Gucci Mane have espoused the virtues of the side hustle.

But by its very definition, the side hustle implies work that occurs in addition to obligatory labor, time, and energy expenditures. Side hustles involve working during off-hours, late nights, weekends, or any time there is free time. A lifestyle focused so much on work comes with risks to physical health and mental well-being. One of the most obvious risks is the fact that side hustles, generally speaking, remain exactly that—side gigs that result in success that is more aspirational than real. Most people never get a chance to turn their passion projects, in this case making games, into full-time gigs. The financial stakes in the new innovation economy are high, but the personal stakes are high too.

A generally overlooked risk associated with the side hustle is unintended health consequences. Whenever I visited a home studio, a coworking space, or some other site for a side hustle, there was usually material evidence of long hours, unhealthy diets, and sleep deprivation—conditions that raise serious questions about young creatives' quality of life. Whether it was stacks of pizza boxes from a late-night work session, trash bins filled with empty energy drink bottles, or someone sneaking a power nap on a sofa, one thing was clear: the grind that comes with the pursuit of the side hustle is not easy, glamorous, or always good for a person's well-being. As a young creative told the *New York Times* in 2016 about her life publishing small independent magazines, "Things look much rosier from the outside than the inside."

Much has been written about the gig economy as a growing number of people are compelled to adopt the side-hustle lifestyle or work main gigs that come with no security, low pay, and no benefits. Optimistic views portray the gig economy as engendering worker freedom, independence, and empowerment. This is certainly the version of the side hustle that massive platforms like Uber or companies who increasingly rely on flexible labor prefer. But narratives like these overlook the immense toll suffered by those who adopt the side-hustle lifestyle out of economic necessity.

Individuals with experience working in triple-A studios frequently complain about the long hours and oppressive working conditions associated with crunch. But indie game developers face their own dilemma—the constant need to use any free time they have to work on their games. When we add up the time required for their paying jobs and their often gratis side gigs, the total number of hours can be well above the forty hours per week we've come to expect. A 2017 report by the IDGA found that independent contract workers often work more hours in game development but earn the least income compared to their regularly employed counterparts.

Concerns about work-life balance came up in several of my conversations with indie game developers, including Jo Lammert and George Royer. "When you are running your own project and you're incredibly personally invested in it, it's very difficult to not do the same thing [work an excessive number of hours]," Lammert admitted. "Because you so desperately want to get something finished, you know, it represents you personally in a way, and I think it can be worse in a lot of ways than working in the mainstream game industry." As she spoke, her facial expression became serious for one of the few times during our conversation.

Young creatives have come of age at a time when our notions about work are changing. For instance, the idea of having one primary job is becoming a thing of the past for many people entering the labor force today. Millennials are much more likely than any other demographic to report working multiple paying jobs. A 2016 report released by CareerBuilder found that 44 percent of persons ages twenty-five to thirty-four work multiple jobs compared to only 19 percent of those age fifty-five and older. One indie game developer told us in a matter-of-fact tone: "Nobody has just one job anymore." The amount of time that young creatives have to devote to their side hustles can be severely limited due to the need to hold down multiple paying jobs.

In addition to the physical cost—lack of sleep and poor diet—that comes with trying to maintain a side hustle, there is the mental cost. What surfaced in many of our conversations was evidence of a heightened degree of stress that comes with trying to create an enterprise without the requisite resources—time, space, money, and staff. The stress level among young creatives is heightened, perhaps, because the separation between work and the rest of life has basically been erased.

Some of the young creatives that I met spoke openly about devoting time and effort specifically to promoting better health and wellness outcomes in the new innovation economy. These efforts employed different types of strategies, but the goal was usually the same: to reset the ruthless work-life imbalance that is emblematic of the side-hustle lifestyle. For some it was yoga, meditation, or some other attempt to self-regulate the mind and body. For others the antidote to physical and mental fatigue associated with the constant pressure to work was making it a priority to spend more time with friends. In a few cases young creatives even organized meetings—formal and informal—to talk openly about the risks associated with side hustling. A growing number of young creatives are electing to advocate for a culture that emphasizes the need for a healthier lifestyle even as they labor to build a better future.

THE EVOLUTION OF CROWDFUNDING

Bootstrapping an indie game means making a game with virtually no money to hire a full team, rent office space, or invest in technology. A few developers we spoke with had turned to crowdfunding to raise the financial capital they needed to make their games. It should come as no surprise that young creatives looking to launch a start-up, social enterprise, or creative project would turn to crowdfunding to generate the financial capital necessary to support their entrepreneurial aspirations. Young people are much less likely than their older counterparts to have substantial financial resources or reserves. This is especially true for a generation—millennials—that has seen its economic prospects take a severe hit as a result of the relatively low wages and underemployment that characterizes their experience in the paid labor force.

The success of crowdfunding is undeniable. Kickstarter, for example, has been the launching pad for more than 425,000 projects since its creation in 2009. Roughly a third, 36 percent, or 153,000 projects have been successfully funded. Near the close of 2018, $3.5 billion had been successfully pledged and fifteen million people had backed a Kickstarter project. The categories that have raised the most money on Kickstarter are games, design, and technology.

There are many elements of disparity in the broader innovation economy, but none may be more significant than the unequal access to the financial capital that fuels entrepreneurial and start-up ecosystems. For example, the most powerful and wealthiest tech companies in the world—Google, Amazon, and Facebook—were all funded by venture capital firms that have clustered along Sand Hill Road, widely recognized as the Wall Street of Silicon Valley. Further, in a study of two innovation hubs, Austin and Seattle, Richard Florida found that the local venture capital firms that provide funding for entrepreneurial enterprises tend to reside and invest in very specific zip codes, thus all but assuring that only certain entrepreneurs will have access to financial capital. Similarly, university-based researchers have access to academic resources, foundations, and federal agencies that fund research and innovation. Needless to say, only a small fraction of the population has access to these sources of financial capital.

It is crowdfunding's promise to democratize access to financial capital that makes it such a laudable idea. Crowdfunding platforms offer hope that a good idea—regardless of its source—can access the financial resources to at least build a prototype to show proof of concept.

My conversation with indie game developers suggests, however, that the culture of crowdfunding is expanding and evolving in ways that has serious implications for its democratizing potential. When I sat down with Jo Lammert and George Royer, their micro-studio had successfully conducted two crowdfunding campaigns, one in 2012 and one in 2014. Crowdfunding changed significantly during that two-year window. "There was definitely a games community on Kickstarter [in 2012], but compared to now it was so much smaller," Lammert told me. She recalled being able to look at all of the Kickstarters for indie games in one afternoon during their first campaign. Two years later they noticed a sharp difference in Kickstarter projects. More specifically, crowdfunding had become more crowded. The rising number of projects made the space more competitive.

. "Yeah, there's no way we could do that now [look at all of the indie game projects on Kickstarter in one afternoon], because there are just so many projects," Royer added.

"Nowadays unless you're a big name already, people expect to see something that's almost done," Lammert said.

"Yes, absolutely. I mean, the Kickstarter market is so saturated now with projects that you really have to have a lot in your hands already to get people excited and interested. Or you have to have a lot of clout in that industry to get people interested in backing it," Royer said.

Many of the young creatives we met understand that it helps to have a rich social network when launching a crowdfunding campaign. Lauren Foster, a young African American entrepreneur, made this point. She and I met one afternoon in a small coffee shop in downtown Austin to talk about her effort to launch a new tech start-up in the food sector. Foster's crowdfunding project was unsuccessful. Reflecting on her experience, Foster compared the start of a crowdfunding campaign to the opening weekend for a film. The films that start strong, she asserted, have a much higher chance of generating momentum and achieving box-office success than those with a weak opening. "Crowdfunding is similar," she told me. Her perspective is backed by research that suggests that people are, in fact, more likely to follow the decision of the crowd and give to a project that others have contributed to. People feel good when they know that a project they have given to is successful. Social scientists call this the "herding effect."

The sheer number of crowdfunding projects these days makes it especially difficult to generate attention. Foster explained that close friends and family—what sociologists refer to as "strong ties"—are crucial to getting a good start and building a foundation for additional support and crowdfunding success. She suggested that entrepreneurs who come from family-and-friend networks that are unfamiliar with crowdfunding or simply do not have money to give are put at a disadvantage in what is an increasingly competitive environment.

Like Foster, many Juegos game developers believe that having a rich social network is crucial to conducting a successful Kickstarter. During the preparations for their crowdfunding project, two indie developers conducted what they called a "pre-Kickstarter campaign." The micro-studio leveraged the relationships they had built with the Juegos collective to begin cultivating a community specifically for their Kickstarter. Weeks before the actual launch, they shared their game with Juegos and started word-of-mouth activity. They invited friends and acquaintances to play-test the game, offer feedback, and share the forthcoming game and Kickstarter campaign with their social networks.

Their approach suggests that having a built-in community that is both familiar with crowdfunding and, in some cases, likely to practice crowdfunding greatly improves the chances for conducting a successful campaign. The indie developers' goal was to raise $10,000. By the end of the campaign they had raised $15,000.

"I'll be curious to see what the space looks like over the next few years, because it is so saturated now that to stand out you have to put so much effort into it," one indie developer told me back in 2015. "Like, you have to have a community already built."

A vital but largely hidden aspect of crowdfunding is the degree to which the model increasingly rewards those who are social- and cultural-capital rich. Social capital refers to the social ties and relationships organizers of crowdfunding campaigns maintain. Cultural capital refers to the reputation and status they accumulate. These aspects of crowdfunding suggest that a model that was designed to democratize access to financial capital is evolving in ways that reproduce some of the same barriers that exist in more elite funding models such as venture capital and university-driven innovation. In the evolving world of crowdfunding, *who you know* and *who you are* may ultimately be more important than having a novel or great idea.

RETHINKING WHAT WE DO WITH GAMES

Today's indie game developers arrived on the creative scene just as games began to ascend to a central role in our everyday lives. Game scholar and designer Ian Bogost writes, "Games have seeped into every aspect of our lives." According to Bogost, this creates an interesting challenge: "What to do with games?" In this shifting cultural landscape, games have become more pervasive and are used in ways that stretch beyond mere recreation.

With the rise of smartphones and artificial intelligence, our world is undergoing steady gamification. The most savvy indie developers respond to both technological and cultural shifts by considering new platforms and new ways of thinking about games.

Members of the Juegos collective made a variety of games. Some developed educational games for clients who viewed games as a literacy platform. Adam Saltsman, one of the few successful indie game developers that we met, made games for brands looking to connect with audiences in more dynamic ways, such as the men's fragrance Old Spice. Saltsman's game for them, like their ads, featured NBA star Dikembe Mutombo. Saltsman even did an iPhone game for the *Hunger Games* movie franchise.

The members of Juegos aspire to make a living making games, but they also want to make games that reflect their creative and aesthetic instincts. They want to produce games that are expressive, quirky, and a rejection of the culture established in triple-A studios. Scholars have used terms such as "venture labor," "aspirational labor," and "hope labor" to describe such modes of work in the digital economy. With maybe one or two exceptions, no one that we met was actually able to make a living making games. As we spent more time with this community, it became clear that the main motivation for bootstrapping was not strictly money or celebrity. No one really believed that they would make games that competed with the titles produced by triple-A studios. That was never the point.

The young developers we met were dissatisfied with the precarious working conditions—contract-based employment arrangements, crunch, and cognitive stratification—that define the triple-A studio system. They were also frustrated with the kinds of games that the industry routinely develops and sells to the world. For them the medium of games is capable of much more, and the games they make differ most from those developed in triple-A studios in terms of the stories they tell and the sensibilities they inspire. Their ultimate innovation, then, has not been games as commercial enterprise but rather games as social enterprise for more imaginative and inclusive storyworlds.

CHAPTER 3

THE SCHOOL OF THE INTERNET

Just-In-Time Learning in the Connected World

Millennials are the most educated generation in US history, and yet stable and meaningful employment for them remains elusive. In 2016 researchers from Stanford University's Equality of Opportunity Project found that young people today are far less likely than previous generations to surpass their parents' earnings and standard of living. More than 9 in 10, or 92 percent, of children born in 1940 earned more than their parents. By comparison, only 50 percent of children born in 1980 went on to earn more than their parents. The study suggests that lower gross domestic product (GDP) and greater inequality in the distribution of economic growth account for the declining percentages of young people faring better than their parents. For many millennials, the promise that each generation will do better than the previous generation—the American Dream—appears to be fading.

In the ongoing quest for some degree of economic security and mobility, young workers change jobs frequently, raising questions about everything from their disposition about work to their reliability. Job-hopping, it turns out, is a relatively common phenomenon among younger workers, past and present. Millennials have been accused of being especially fickle, unreliable, and much more likely to change jobs than previous generations of young workers. However, a look at the data suggests that the notion that millennials change jobs more frequently than Generation Xers is simply not true.

A 2017 report by the Pew Research Center found that workers ages eighteen to thirty-four are just as likely to stick with their jobs as their counterparts from Generation X at similar ages. Examining US Department of

Labor data, Pew found that 63 percent of millennials (in 2016) compared to 60 percent of Generation Xers (in 2000) had been with their current employer thirteen months or more. Further, Pew reports that millennials were just as likely as Generation Xers to have been in their job for five years.

Economists note that it is actually normal and even good for the economy for younger individuals to change jobs more frequently than their adult counterparts. In one study a team of economists calls this behavior "occupational fit." Because young workers are just starting out in their careers, they are much more likely than older workers to sample occupations in order to find those in which they are most productive. As a result, the authors write, "Young workers . . . spend more time in transition between occupations."

When young people change jobs, they are not being fickle; they are actually being rational. More specifically, they are looking for jobs that represent a better match for them and, usually, better compensation. Henry Sui, a professor at the Vancouver School of Economics, told the *Atlantic*, "People who switch jobs more frequently early in their careers tend to have higher wages and incomes in their prime-working years." Sui adds that when it comes to practicing occupational fit millennials are like previous generations. However, he suggests that millennials approach to work differs from people their age in generations past in one important aspect. More specifically, the economist explains that while millennials are not changing jobs at rates higher than previous generations (separation rate), they are more likely to try an entirely new job (occupational switching). Writing about the growing phenomenon of occupational switching for the same *Atlantic* article, Derek Thompson asserts, "Young people aren't quitting more. They're experimenting more."

Many of the young people that we met in our fieldwork were certainly experimenting with their careers. The overwhelming majority of these young people were either going through or anticipating going through what some of them called a "career pivot." Trapped in jobs that rarely promise occupational mobility or financial stability, many millennials face a need at some point to alter the trajectory of their career path. This typically requires an occupational makeover of some kind.

The mode of career pivot that I frequently observed is not simply about finding a better-paying job. It is also about asserting greater control over one's occupational identity. Many young creatives looking to make a career change told us that they simply wanted to find work that was not only financially rewarding but personally fulfilling too. In other words, occupational fit for them is as much about quality of life as financial status. It speaks to the desire among a surging number of young creatives to build an innovation economy that prioritizes making an impact just as much as making money. This attitude is reflected in the generational mantra "Doing well by doing good."

THE CAREER PIVOT

The career pivot involves shifting to a new line of work. This move often requires growing an individual's human capital, which, among many other things, means learning new things and developing new skills. Whereas in the past, employers might have invested in professional support for their employees through, for example, workplace mentorship programs, professional development opportunities, or further education, young workers today must usually take it upon themselves to enhance their job skills. This reflects the rise of what some call the "risk economy." In the world of work that young creatives navigate, individuals are increasingly expected to take on a greater share of the responsibility for upgrading their workplace skills and advancing in their careers. As more and more employers retreat from long-term financial commitments to their employees, they also retreat from paying for programs and infrastructure that help their employees build the necessary skills to stay competitive in an ever-changing labor market.

Traditionally, school has been the primary place where young people develop human capital. We generally think of school as the place where young people go to learn things, including the kinds of things that translate into opportunities to earn income. Academic skills—including literacy and numeracy—and vocational skills—including the knowledge required in specific trades—presumably prepare students for the workplace. But public schooling in America over the last few decades has experienced a crisis defined by a growing discrepancy between the skills schools develop and the skills the new economy demands. As the economy continues trending toward what some experts call "skill-biased technical change," the kinds of skills that are increasingly valued in the new economy—expert thinking, design, problem-solving, data literacy, and computer-technical ability—are precisely the skills that most schools struggle to cultivate. Among all of the criticisms of schools that circulate today, none may be more powerful than the claim that they are simply not designed to help the majority of young people develop the skills that are valued in a rapidly evolving knowledge economy.

The crisis in public education notwithstanding, additional schooling is not always an option for young people looking to make a career pivot. Some of the young creatives we met had already earned bachelor's degrees, which for many included an accumulation of financial debt. Others were in no position financially to pursue a postsecondary credential. If formal education is not an option during a career pivot, where do young creatives go to learn the things they need to know to switch to occupations that are more secure, creative, or rewarding? Many of the young creatives we met sought alternative places to upgrade their skills or develop a whole new set of skills necessary to engineer a career pivot.

THE CLASSROOMS OF TOMORROW

A growing number of young creatives are enrolling in what I call "twenty-first-century skills-building accelerators." These are entrepreneur-driven entities such as coding academies and design schools that offer boutique-style hands-on curricula over a fixed period of time and that teach students relevant jobs skills, such as coding, design thinking, and interaction design. Many innovation hubs—San Francisco, Seattle, Boston, Austin—are prime locations for these kinds of academies that help young aspirational techies and creatives pivot with confidence.

Our research team spent some time in a design school that offers a nine-month course in interaction design. Most of the students we encountered there were in their middle to late twenties, the age when the urge to pivot or the desire to assert greater control over one's career trajectory becomes more salient. All the students had bachelor's degrees but were dissatisfied with their work experience and future prospects. They were not just looking to change jobs, they were looking to change their lives.

One day while I was visiting with the founder of the design school, he told me that teaching the students the skills in the curriculum like participatory design, user testing, and wire framing was not especially difficult. The greater challenge from his perspective was helping students develop the confidence and the autonomy crucial in today's economy. They were all looking to make a career change, and that required a different kind of disposition or, as he put it, "the confidence to say, 'I have the skillset, the aptitude, and the intellect to get over there even if it's really hard.'"

Unfortunately, the tuition for these twenty-first-century skills-building accelerators can reach as high as $20,000 to $30,000, setting the cost outside the scope of possibility for the overwhelming majority of young people looking to switch to a high demand occupation. Young creatives have, in response, flocked to a fascinating and vitally more affordable "place" with the goal of growing their human capital: the internet. Young creatives' adaptive use of the internet is radically transforming how they learn. Equally important, this practice expands the geography of innovation.

Wiley Wiggins, the bootstrapping indie game developer, is a prime example of this practice in action. Because they lack money, indie game developers have to wear many hats and learn new things as they go. Where do indie developers go to learn new skills related to making games? Wiggins and several of the other developers that we met noted how they leveraged the internet to grow their knowledge and expertise. This illuminates another feature of tech ingenuity that is generally overlooked by the tech addiction narratives that tend to stigmatize young creatives and their relationship to technology. Whereas educators, policy makers, and parents tend to view the internet as a deterrent to learning, young people tend to see the internet as an asset for learning. For them the internet is a place where knowledge is distributed in ways that allow access to whole new worlds. Over the years I have observed repeated instances of young people turning the internet into the world's biggest and most networked classroom. This was certainly the case for Wiggins and many of his peers.

Wiggins is one of the more interesting young creatives that I met during my research. In addition to cofounding the indie game collective Juegos Rancheros, he helped to organize a local indie game festival that is gaining in reputation and attention among indie game developers across the US and other parts of the world while also attracting sponsorship dollars from companies like Sony. His side hustle also includes running a micro-studio to build indie games. Wiggins's main gig involves doing UX design for a software company. With a resume like his, I was stunned to learn that he had not attended college.

"It's a little embarrassing when I talk to people about, you know, high-minded stuff, and they're like, 'What school did you go to?' And I'm like, 'Well . . . ,'" he acknowledged with a laugh. "But you get over it. I don't think there's any right way to learn career-ready skills," he said in a moment of personal reflection. "My parents could not afford to send me to college."

During a lengthy conversation with one of our researchers, Wiggins explained his decision not to attend college this way: "It just didn't really seem like an option. There wasn't savings, and my parents could not afford to send me to college. I wasn't a good enough student to get scholarships, and I don't think that I would've been a good loan candidate." In addition, Wiggins felt alienated from high school, bored by a curriculum he believed was a fast track to nowhere. The dumbed-down curriculum, ineffective approaches to technology, and misguided education techniques simply did not inspire him. "I didn't feel engaged in any of the learning that was going on in high school. All of the computer stuff that I was learning—the stuff that I was actually interested in—was happening outside of class," he confessed.

The fact that he did not go to college meant that much of what he learned about UX design (his paid gig) and game development (his side gig) had to have happened somewhere else. For Wiggins, that somewhere else included the internet. For Wiggins, online forums, online gaming communities, and university-based open courses were lively places to learn about the many things that interested him, including graphic arts, game development, and programming. Wiggins's extracurricular learning had two things that his formal education lacked—passion and purpose. Wiggins knew precisely how he would apply the knowledge he sought online. He wanted to make interactive media and games.

For instance, he began frequenting TIGForums, the world's biggest online forum devoted solely to independent game development, to grow his knowledge about game creation. The online forum was a great place to learn about specific aspects of game development, including design, art, and sound, from people who were actively developing games. Wiggins used the internet as a tool to fashion a learning ecology in which he had direct access to knowledge and expertise related specifically to game design. The learning in internet settings tends to be hands-on, experiential, and connected to a tangible interest or goal. Learning in school is most often theoretical; learning via the internet is practical.

Wiggins joked that instead of going to college he went to "the school of the internet." This was more than a sly quip. It was also a reference to a style of learning that is familiar to many young people looking to spark an interest in a subject or a career path that most schools neglect.

CONNECTED LEARNING

A group of researchers that I collaborated with on a separate project commissioned by the MacArthur Foundation refer to this style of learning via the internet as "connected learning." The idea of connected learning recognizes that the informal circulation of knowledge via online platforms and networked communities is radically transforming the way many young people learn today. This dynamic form of learning is powered by interests, passion, peers, technology, action, vibrant social networks, and a bias toward making knowledge actionable in ways that school rarely does. Connected learning is not simply about sitting in front of a screen. Rather, connected learning suggests that learning is more effective when it happens across multiple settings, such as in school and out-of-school, with teachers and with peers, offline and online.

Learning in the context of the internet is very different than learning in schools. Whereas learning in schools is top-down and teacher-driven, learning via the internet is bottom-up and peer-driven. If learning in school is restricted to the space between four walls, learning via the internet is networked and typically happens beyond the confines of a physical setting. Also, in schools, learning is typically rote and inactive whereas learning via the connected world is dynamic and experiential. Finally, learning in school is typically arbitrary and disconnected from the real world. By contrast, learning with the internet is intentional and connected to real-world aspirations.

Wiggins and his peers provide clear evidence of the spread of connected learning in the new innovation economy. It was quite common for the indie game developers we met from Juegos Rancheros to turn to online forums and networked communities to learn something they needed to know related to game development. This strategy used within the world of connected learning is sometimes referred to as "on-demand" or "just-in-time" learning, and it typifies the way a growing number of young creatives build their human capital.

These online communities and forums "are springing up all the time," Wiggins told us. The key in his mind to fully leveraging online forums and just-in-time learning is "knowing what you do not know." Wiggins elaborated during a lively conversation with a member of our research team: "Asking the right questions," he explained, "is more important than having a predefined list of where to go. It seems like the big struggle is figuring out what the names for the things that you're trying to learn are. If you don't have that, then it can be very difficult."

The internet-enabled learning that Wiggins described to us was much more creative and complex than a Google search or watching a video tutorial. More specifically, Wiggins's school of the internet makes learning social and actionable. For instance, it requires searching, finding, and learning how to participate in the online communities and social networks through which particular knowledge and information flows. Additionally, the school of the internet is not about sitting in front of a computer and passively consuming information. Rather, it provides access to vibrant conversations, the exchange of ideas and information, tools for problem-solving, and the cultivation of knowledge and social networks crucial to executing a specific project.

In his adoption of the internet as a learning tool, Wiggins also turned to a new generation of open courses that many universities offer. "I followed along with ITP [Interactive Telecommunications Program] at NYU and learned so much just from them making a lot of the courseware open. I've written to professors at ITP and they've totally helped me with stuff," Wiggins said. The open university courses did more than provide access to useful information. Wiggins also used the courses as an opportunity to diversify his social network. Through the course he met professors and other students that he could tap for insight.

Wiggins acknowledged, however, that there are some limits to the school of the internet. "Serious work on things, like psychophysics, stuff like that, you know, signal processing—it's all going to be textbooks," he said. "And they're not things that you can just walk in and read and digest without a professor or some sort of support group to help you with that." To gain access to university textbooks, Wiggins turned to his local social network, often relying on friends who attended the University of Texas to access to resources that were only available in the library. Wiggins continued, "Like, if you're doing hard science, you kind of need to go to school, and I do have an interest in hard science. So that has always been a thing that's kind of been lacking a little bit."

As Adam Saltsman began to develop his interest in making games, he also turned to the school of the internet. Saltsman said, "It's great if you can have an in-person teacher," but aspiring indie game developers seldom have access to that kind of instruction and expertise. Saltsman noted that the next best alternative is finding "an online community where you don't post your artwork because you want praise and support but you want people who are smarter than you and have more experience than you to tell you how to make it better. This was really important for me because I didn't have a lot of access to in-person teachers." He credits these online communities with cultivating the human capital—knowledge—and social capital—interpersonal connections—that made it possible for him to earn a living making games.

But Saltsman also offered an important caveat about the school of the internet by acknowledging that access to and successful participation in on-line forums is easier for some than it is for others. "I know that basically as a white guy it's a little bit easier to be accepted into a lot of online communities, and it's a little bit less risky to participate in those things," he told a member of our research team. By contrast, women and African Americans, for example, may be exposed to the toxic behavior associated with life in the geeky, male-dominated, and often exclusive world of tech. The school of the internet can be inhospitable to anyone who does not fit the "tech bro" type.

Throughout my research over the years, one of the things that has stood out is the extent to which the internet functions as a learning laboratory for young people, a place they go to learn the things they want to know. In many cases the things they are passionate about are not necessarily or effectively taught in school. This is especially true for students who attend resource-constrained schools. I have met students who have used the internet to learn how to play an instrument, edit a digital video, or make a gaming computer. This explains, at least in part, the popularity of YouTube, a robust ecosystem of learning, among young people.

YouTube tutorial videos do not just *tell* viewers what to do; they *show* us how to do it. This form of knowledge transmission and knowledge acquisition underscores, among other things, just how out of touch most schools are. When young people watch a YouTube tutorial, they are not merely interested in learning something, they are interested in doing something. Their learning is self-directed for practical purposes. Similarly, online forums and communities offer a place to connect with others with a purpose in mind, to ask and answer questions, and to develop real-world and real-time problem-solving strategies.

There are many skills that the new world of work requires, and just-in-time learning—as well as other skills displayed by Wiggins and many other young creatives—is certainly among them. In today's economy workers have to be flexible earners *and* flexible learners. What I find striking about young creatives like Wiggins is that, in the pursuit of their creative aspirations, they are expanding our notions of how and where learning happens. They are not only learning a new skill, such as coding or interaction design. They are also identifying the best resources, networks, and methods for learning how to write code or practice interaction design. Learning a new skill is crucial to initiating a career pivot. Also crucial is identifying the resources and places that strengthen your ability to learn new things. In other words, young creatives are doing more than just learning. The tech ingenuity they practice suggests that they are also learning how best to learn.

HOW COLLEGES ARE FAILING YOUNG CREATIVES

Wiggins's ingenious manipulation of online communities and courses to engineer whole new learning environments is a method that he and young people like him are steadily improving. Once they identify what they need to know, they figure out ways to learn it, in part by leveraging the crowd-sourced, community-driven, and open platforms that live on the internet. For them, the internet is an essential resource in the process of growing their human capital and pivoting to another career.

The real dilemma that young creatives like Wiggins face is a faltering education system. Wiggins's experience underscores the stinging criticism that schools are simply out of touch with the modern world. In his book *The Case Against Education: Why the Education System Is a Waste of Time and Money*, Bryan Caplan, a George Mason University economics professor, argues that the "college for all" mantra that dominates education, policy, and public discourse should be abandoned. Colleges, Caplan claims, teach subjects that are largely irrelevant in the modern jobs economy. He asks, "Why do English classes focus on literature and poetry instead of business and technical writing? Why do advanced-math classes bother with proofs almost no student can follow?" Caplan believes that students are on to something when they complain that school is irrelevant because much of what they are required to learn they will never use in the real world.

Caplan does not argue that college has no benefits. Instead, he argues that college does not do what most people assume—teach people skills that are in high demand in a rapidly changing jobs economy. The real benefit of college, according to critics like Caplan, is the role it plays in "educational signaling." Caplan writes, "The labor market doesn't pay you for the useless subjects you master; it pays you for the preexisting traits you signal by mastering them." In other words, college graduates do not necessarily get jobs because they have mastered highly sought-after skills. With the exception of a few majors, such as computer science, statistics, and nursing, students seldom learn skills that directly translate into the employment sector. Rather, they are hired because finishing college signals to employers that they can likely do things such as comprehend instructions, follow rules, and show up when expected.

Wiggins does not believe that college would have made a big difference in terms of his career. "I'd probably have the same job now that I would've had if I had gone and studied interaction design in school," he said. His experience notwithstanding, the data consistently supports what some experts call a "college-wage premium." Today, individuals with a college degree enjoy significantly higher lifetime earnings than those without a college degree. A study by the Pew Research Center found that the only thing more expensive than going to college is not going to college. According to Pew, the income gap between a college graduate and someone with just a high school diploma continues to grow. In 1965 the average income gap between a college and high school graduate was $7.49. Four decades later, in 2014, the gap was $17,500.

The story of young creatives like Wiggins suggests that in our rapidly evolving economy, just-in-time learning is becoming the norm. People entering the workforce today are predicted to have several jobs over the course of a lifetime. As a result, a surging number of workers will constantly need to build their skills simply to remain employed. In the meantime, colleges would do well to rethink the learning and career–oriented trajectories they recommend to their students, since their current configurations do not reflect knowledge needed in the real world.

Like a growing percentage of young people, Wiggins decided that he could not afford to pay the substantial college tuition for a degree that would signal to potential employers that he was reliable but fail to adequately equip him with the skills to nimbly pursue work in an economy biased toward higher-order thinking and doing skills. "College is expensive," Wiggins said. "Like, if you go to college and get a degree in something where you're, like, 'Oh, why did I even do that?' I mean, yes, you've got a piece of paper that'll tell an employer that you'll show up to work, and you can be counted on to do stuff. But you don't need a $90,000 piece of paper to do that."

Wiggins decided instead to invest his resources—time and energy—in identifying and cultivating the knowledge and skills that would propel him along his preferred career path. Rather than incur substantial financial debt for a credential—a bachelor's degree—that did not guarantee a pathway to the career he coveted, Wiggins fashioned his own learning ecology in which to grow the human capital that was most relevant to his ambitions to design interactive media.

Among the many generational divides related to the rising influence of technology in our lives, none may be more revealing than the different views young people and their older counterparts have about the internet as a viable learning platform. Whereas educators and policy makers view the social web as a deterrent to learning, many young people believe it is a complement to learning. For a growing number of young creatives, the internet is a vital space for investing in just-in-time learning to enhance their human capital. In the world of tomorrow connected learning is a key requirement for young creatives looking to make a career pivot and change the trajectory of their lives.

CHAPTER 4

HUSTLE AND POST

Hip Hop, Social Media, and Pop Music Innovation

Hip hop has a legacy of innovation that dates back to its origins in the place that KRS-One famously called the "Boogie Down," 1970s Bronx, New York. Many of the characteristics that I associate with the new innovation economy, including tech and social ingenuity, the side hustle, DIY ethos, grit, and experimentation have been a way of life in hip hop since the beginning. The culture of hip hop has thrived over the last forty-plus years, in part, by mastering the side hustle and expanding the geography of innovation.

Many of the pioneering DJs in hip hop, such as DJ Kool Herc and Grandmaster Flash, used public parks and community centers as their dance clubs and laboratories for crafting techniques that transformed the role of the DJ in global music culture. Hip-hop visual artists used public buildings and subways as their studios to make graffiti art and a name for themselves. Sylvia Robinson recorded rap music's first commercial hit, "Rappers Delight," in a nondistinct studio for $750. With virtually no access to retail distribution, young hungry rappers used store parking lots and the trunks of their cars to sell their mixtapes on cassettes and, later, CDs. Today we call these improvisational retail spaces pop-up shops.

In the early to middle 2000s, Public Enemy's Chuck D was one of the first to understand that the internet was a space that offered artists creative independence, access to fans, distribution, and the resources for building a community around their music. The social media practices developed by hip-hop artists, such as the use of social media to talk directly with fans, release music, sell merchandise, and tell their brand story, are standard music industry practices today. For all of the attention hip hop has received over the years for the bold music and colorful characters it has spawned, the real story is how virtually everyone who has made a name for him- or herself in hip hop's sprawling economy has mastered the skills necessary to succeed in the new innovation economy.

It likely comes as no great surprise that the internet plays a prominent role in the rise of the new innovation economy. More surprising is how young creatives have populated the internet, using it as a space to mobilize some of the critical assets necessary to pursue their artistic, economic, and civic ambitions. Once regarded for its ability to provide unprecedented access to the world's information, the internet also offers access to particular communities of knowledge and expertise, serves as a hub for content creation and circulation, and mimics some of the core functions of innovation labs by fostering ideation, experimentation, and iteration. For many young creatives the internet has become a dynamic hub for developing the social networks and inventive practices that drive the new innovation economy.

Most young aspiring music artists do not have access to studios, record labels, or marketing professionals. For them, the internet has become the ultimate platform for hacking the barriers that make it difficult to break into the music and entertainment business. Certain aspects of a changing tech landscape—laptops loaded with music-making software, online video channels, and music-oriented mobile apps—lower the barrier to entry in the music business. The savvy exploitation of these technologies accelerates the rise of new disruptive styles of music by artists who make up for their lack of corporate assets with an abundance of creative swag and social media savvy.

In the era of do-it-yourself music production and distribution, independent artists face many obstacles to sustaining a livelihood from their creative labor. Independence is a double-edged sword. On the one hand, artists can use a mix of platforms and channels to produce, promote, and profit from their creative labor. On the other hand, artists take on the risks that were once primarily shouldered by music industry personnel, such as artists and repertoire representatives, producers, and marketers. With the new tech platforms available today, DIY music is more possible, but that does not mean succeeding in the music business is more likely. Technology lowers the barrier to entry for all, thus intensifying the competition for attention in a culture in which virtually everyone is a media producer.

If artists are willing to work around the clock making and marketing their music, they can sometimes make a living from their art. Perhaps more so than any other creative worker, musicians struggle with the possibilities and the perils of freelance labor and technological change. These struggles may be a common feature in the growing gig economy today, but they have long been enduring features of life for music artists. Musicians are among the longest tenured class of gig workers in America's economy.

SOUNDCLOUD: CREATING A YOUTUBE FOR AUDIO

In 2007 two young Swedes named Alexander Ljung and Eric Wahlforss met at the Royal Institute of Technology in Stockholm. According to Ljung, "We were at the university to get time to build crazy projects." Shortly after meeting they discovered that they both had a love for music. Ljung, a sound designer, had built a home studio to make music. In 2011 he told *Fast Company*, "I made some terrible music, but it was pretty well produced." Wahlforss was an electronic musician. After a number of conversations about possible projects they turned their attention to sound on the web. They began thinking about building a social media platform that served the producers of sound. "Everybody doing music on the web only focused on the consumer side of things," Ljung told *Fast Company*.

At that time Flickr and YouTube were generating massive headlines and web traffic for their role in the making of a new kind of internet—one that encouraged content creation rather than content consumption. Flickr was a service for those who created photos. YouTube served those who created videos. When Ljung and Wahlforss wanted to share their original music with their peers, there was no easy way to do so. "It was just really, really annoying for us to collaborate with people on music," Ljung told *Wired* in 2009. "I mean simple collaboration, just sending tracks to other people in a private setting, getting some feedback from them, and having a conversation about that piece of music."

Flickr made it possible for photographers to connect via the social web. And YouTube did something similar for those who uploaded videos to the site. "We didn't have that kind of platform for our music," Ljung said. This, as entrepreneurs like to say, was their opportunity space. According to Ljung, "We decided to focus on the creator side of sound and make use of all the cool things that were happening on the social web."

They moved to Berlin and began coding a platform they believed would be useful to people who created sound. Berlin was emerging as a hot start-up capital, Europe's version of Silicon Valley. Ljung and Wahlforss called the new product they were building SoundCloud. They pitched the service in very clear terms: "SoundCloud is a sharing platform for sound." The site was officially launched in 2008.

Once SoundCloud launched, the founders faced the problem that all new internet entrepreneurs face: finding users. They reached out to local DJs in Germany's electronic music scene and encouraged them to use SoundCloud to share their music. In addition to sharing their music with others, the DJs used SoundCloud to build their own online community. Soon both aspiring and established music artists began using the service. Looking to take their careers in new creative directions, Moby and Beck used SoundCloud to release fresh mixes. But it was the indie and aspiring artists who really grasped the promise of SoundCloud. These early adopters liked the idea of a platform that fostered community, collaboration, and feedback from colleagues and a devoted listening public.

Less than a year into its launch SoundCloud was threatening the dominance of MySpace in the corner of the internet that catered to music fans, trend followers, and underground artists. Though the once-popular social network site's role in the new media universe had been significantly diminished with the rise of Facebook, YouTube, and Twitter, MySpace was still a popular site for musicians looking to build a community around their music. Artists were drawn to SoundCloud, in part, because it offered a great opportunity to connect with other artists and share ideas about music making. By 2012 SoundCloud was emerging as a go-to destination used by musicians, corporate media, and podcasters. Reputable news organizations shared their stories on SoundCloud. The White House, under President Barack Obama, began using the platform to share speeches and other communication. The SoundCloud publicity machine framed the technology as the premier service for what was new, now, and next in the world of sound.

REMIXING HIP HOP'S DIGITAL UNDERGROUND

A wide community of young artists, podcasters, DJs, and audio producers turned SoundCloud into a viable hub for audio creation, circulation, and curation even as the founders struggled to monetize the swirl of activity the platform sparked. No creative community used SoundCloud more effectively than a new generation of hip-hop artists who adopted the platform to work around the industry barriers that exclude them. Some of the aspiring hip-hop artists we met during our fieldwork mentioned how they use SoundCloud to connect to other hip-hop artists around the US and to expand their music-related social network. They use SoundCloud to elicit feedback on a new song, identify potential collaborators, create a portfolio of content, share their music, and pursue potential gigs.

SoundCloud is the platform of choice for a large number of underground hip-hop artists. There are several reasons for this. First, it is easy to set up and begin using SoundCloud. Sign up for a free account, and within minutes you could be posting songs that the global internet population can access. Second, the designers made it easy to embed SoundCloud audio across the social media landscape. A song posted on SoundCloud generates a URL that can be shared on Twitter, Facebook, or Instagram in the blink of an eye. Third, SoundCloud connects seamlessly to the smartphone lifestyle many young creatives live.

SoundCloud, with the help of many young creatives, emerged as a fertile space in hip hop's underground. Throughout the history of hip hop, aspiring artists have used the underground terrain to invent new forms of expression in music, fashion, art, dance, film, and design that eventually spread into mainstream culture. The use of SoundCloud by rappers, DJs, and beat makers follows a similar trajectory. This was especially true in the rise of what became known as SoundCloud rap.

SOUNDCLOUD RAP BLOWS UP

Among all the rappers and subgenres of rap to establish a presence on Sound-Cloud none has been more influential than a generation of artists who earned the moniker "SoundCloud rappers." The term is a reference to a subgroup of rappers, many in their teens and early twenties, who have built an identity and a community via SoundCloud. In addition, these artists have helped craft a style of rap music that is irresistible to some and irritating to others.

Most observers characterize SoundCloud rap as low-fi, a reference to its deliberately uncorporate demeanor. SoundCloud rap features a thick distorted bass laced with lyrics that are frequently incendiary and unhinged. In SoundCloud rap you might hear songs about murder, suicidal thoughts, sex, or teen angst. Some even liken the hyper-aggressive posturing in Sound-Cloud rap to the punk movement. The roughly two-to-three-minute tracks are crafted for young streaming music consumers who have found their way to this particular corner of hip hop's digital underground.

In comparison to corporate rap—think Drake—which has become pretentious and predictable, SoundCloud rappers bring a very different disposition to the rap game. Whereas corporate rap is polished, overproduced, and carefully calibrated for mainstream viability, SoundCloud rap has a reputation for being unpolished, underproduced, and carefully calibrated for street credibility.

The incendiary lyrics in SoundCloud rap are certainly not new to hip hop. In fact, there is a long tradition of lyrical insurgency and absurdity in hip hop often prompted by a feeling among young MCs on the come up that the establishment—namely, major label rappers—abandon what originally made rap relevant, such as bold innovation, a say-anything orientation, and sparse instrumentation. SoundCloud rap is part of a lineage of rebellious rap that promises to stay true to hip hop's legacy of insubordination, creative independence, and a "keeping it street" ethos. A 2017 *New York Times* profile of key figures in SoundCloud rap called the subgenre "the most vital and disruptive new movement in hip hop."

SoundCloud rap has become notorious in part because of the "don't give a f@#k" lifestyle that some of its most well-known artists embrace. Performers like Lil Purp, Lil Peep, Smokepurpp, XXXTentacion, and Trippie Redd play their roles as cultural provocateurs with assured consistency. In the case of XXXTentacion, the outlaw sensibilities common in SoundCloud rap were played out to their most extreme end when the rapper was murdered in June 2018. The trappings of SoundCloud rap—everything from the brightly colored dreadlocks, tatted-up faces, in-your-face lyrics, and penchant for Xanax popping—inject a heavy dose of bombast into the movement. But for all of the amped-up and even deadly theatrics that characterize SoundCloud rap, its ties to hip hop's legacy of innovation are undeniable. Like generations of hip-hop artists before them, SoundCloud rappers turned an unlikely platform, in their case social media, into a vibrant space of innovation and economic opportunity.

SoundCloud rappers are linked to a lineage of DIY music artists who made the internet their primary hub for pop culture innovation. One of the early masterminds in hip hop, Chuck D, helped establish the internet as a viable space for hip-hop creatives when, in 2000, he embraced the internet and the future of music even as the industry and popular bands like Metallica resisted. During MySpace's brief window of popularity, Drake used the internet to manufacture buzz and boost his commercial ascent. Many young rappers in the social media era credit the blueprint developed by Los Angeles hip-hop collective Odd Future for demonstrating how the internet could be leveraged to build a career in music without any support from a label.

Odd Future engaged the internet to launch a brand of pop music that was fiercely independent, releasing their first album in 2008. Using Tumblr as their creative hub, Odd Future released a string of albums for free. One music industry publication characterized their music as having "tightly penned raps, sonic cohesion and thoughtfully executed conceptual arcs." The visual sensibilities in Odd Future's homemade music videos were calibrated for maximum shock value and reflected the youthful angst associated with growing up in a tech-rich but precarious society. Odd Future made its music in this generation's most reliable studio—the laptop computer. Without any help from industry gatekeepers, Odd Future launched the careers of talented artists Tyler, the Creator; Syd the Kid; and Frank Ocean. In a 2011 cover story about Odd Future's inventive deployment of social media, blogs, viral videos, and streaming, *Billboard* magazine noted that the group "may just be the future of the music business."

Odd Future leveraged their internet fame to essentially act like a miniature media conglomerate. In addition to their music, Odd Future launched a sketch comedy show on Cartoon Network's *Adult Swim*; a commercial-free online radio station; a line of street fashion; and a boutique store where you could buy their merchandise. On Odd Future's heels, other young creatives in the music industry have realized that the days of corporate mega-advances and sweet contracts represent a bygone era, and they are using DIY media and merchandising as a way to expand their creative and earning potential. None of this, of course, would be possible without the internet as a crucial space that expands the geography of innovation.

The popular and music press struggled to understand Odd Future. Nothing about them resembled anything about hip hop that was familiar. Their videos drew as much from the horror film genre as they did any Black visual aesthetic. The group's fashion style seemed influenced more by skater culture than Black culture, and the audience for their music was primarily young white teens with punk rock sensibilities. However, in terms of their ingenuity, Odd Future was pure hip hop. Like earlier generations of young hip-hop creatives, Odd Future exploited social and technological shifts to launch their unique style of pop music. They transformed marginalization into a source of innovation.

Odd Future's long-term impact is not how they mastered the art of making music without the support of a corporate studio. Rather, their impact is in how they mastered the possibilities of the connected world by choosing the internet as their innovation hub, a space, that is, to execute their brand of musical creativity and business acumen.

SOCIAL MEDIA AND THE HIP-HOP INNOVATION ECONOMY

With nothing but their tech, ambition, and hustle ethos, SoundCloud rappers work within real financial and technological constraints. Part of their tech and social ingenuity involves setting up shop in modest home studios and relying on sparse beats. Discussing the process he uses to make music, Ronny J, a SoundCloud rap producer, told the *New York Times*, "Everything is literally made with me not even getting out of my bed and a kid coming and getting on the mike and screaming or rapping." Ronny J's comment reflects a degree of posturing that has long played a strategic role in hip hop. Throughout the years hip-hop artists have suggested that their art is simply a matter of keeping it real. But hip-hop artists possess a savvy understanding of the entertainment marketplace—and what teens want in particular. It would be a mistake to read Ronny J's comment literally and, thus, dismiss the creators of SoundCloud rap as indifferent to the craft of music making.

SoundCloud rappers are not bedroom artists because they are lazy about music. Rather, they make their bedrooms and laptops their studios because they have nowhere else to record. The unabashedly in-your-face demeanor of SoundCloud rap can be attributed, in part, to the fact that its pioneers were unencumbered by music executives who gauge every track, lyric, or image for financial impact. Historically, that kind of freedom from corporate dictates—no rules, no bosses, and no commercial constraints—has been a starting point for bold innovation and disruption in hip hop.

SoundCloud rappers embody the venerable practice of tech ingenuity in hip hop. Like previous generations of hip-hop practitioners, they use technology to craft their creative identities, expand their art-making opportunities, and pursue their entrepreneurial ambitions. They rely on a hodgepodge of tech tools to create and circulate their music, including laptops, digital audio software, smartphones, mobile apps, and, of course, social media. The most popular SoundCloud rappers are a fixture in the youth-driven social media ecosystem made possible by Twitter, Instagram, and Snapchat. These MCs cultivate hundreds of thousands of social media followers, who not only listen to their music but also help generate online buzz, wide visibility, and pop credibility.

Like many of the young creatives I met throughout my research, Sound-Cloud rappers use social media in a dynamic and strategic fashion. They routinely produce tweets, Instagrams, and snaps that go viral, build a loyal fan base, and generate publicity without the benefit of a corporate-driven hype machine. Through social media, they talk directly to fans, establishing a connection that is at once personal and public. In a 2017 interview with the *New York Times* Smokepurpp, a popular SoundCloud rapper, said that before his music career took off with touring and recording he spent 85 percent of his time working on his image and 15 percent of his time working on his music. The image work was largely time spent using social media to build an identity and a community around his music.

"This generation and wave of rap are using social media in a way that no other class has," said Cole Bennett, a music video director, to *Rolling Stone* magazine. The secret to their social media success is no secret. They use social media the way that most teens use it—every day and throughout the day. According to Bennett, "Mainstream acts are only going to make a tweet when they drop an album and they may not tweet for months. These young guys are talking to kids every day." A 2017 survey that I conducted with NORC, a nonpartisan research group at the University of Chicago, corroborates this claim.

The survey suggests that young people use social media for nearly everything: to connect with friends, family, and acquaintances; to consume news, information, and entertainment; to express their politics and participate in civic life. But young people also use social media to navigate a precarious economy and to compensate for social immobility—especially young people who are lower-income, African American, and lack a post-secondary credential. For many young people who are hustling to build their own futures, social media is more than just a tool for casual connectivity; it is also a tool for strategic connectivity and in some cases economic opportunity.

If you look past the outlandish image and belligerent lyrics that characterize parts of the SoundCloud rap movement, you may notice how the young creatives use social media strategically to craft a persona, cultivate a fan base, gauge public response to their music, and generate paid gigs and merchandizing opportunities. SoundCloud rappers leverage the tech tools of their day to do what those on the margins of society have always done: make something out of nothing.

TIDAL WAVE: THE STREAMING MUSIC ECOSYSTEM

In 2017 SoundCloud rap began to gain substantial traction and attention. The rise of the movement signaled significant behavioral shifts in the consumption of music and, more specifically, the emergence of streaming as the platform of choice for a majority of young music consumers. The new pop music landscape—DIY production, social media, and streaming—is driving the future of music and forcing the corporate labels to change their ways.

Hip-hop creatives saw the future of music before most in the music business. Chuck D embraced internet distribution before the infrastructure for streaming—broadband connectivity, smartphones, mobile apps—was built. In 2012 hip-hop icon Dr. Dre and his business partner, Jimmy Iovine, began building Beats Music into a streaming music platform that anticipated new modes of music consumption driven by the widespread use of mobile apps, high-end headphones, and algorithmic recommendations. In 2014 Beats was acquired by Apple to help reinvent its streaming music service, Apple Music. Hip-hop legend and Hustler-in-Chief Jay-Z also had his eye on the future when he unveiled the streaming platform Tidal in 2014.

In November 2014 *Billboard* magazine, the music industry's leading trade publication, announced that it was transitioning "from a pure sales-based ranking to one measuring multi-metric consumption." The new algorithmic measurement system recognized the obvious: streaming is the future of music consumption. The algorithm, developed by Nielsen Entertainment, included on-demand streaming and digital track sales, so that *Billboard* could measure what it called "consumption activity." Under the new system, for example, 1,500 song streams from an album were equivalent to one album sale.

The announcement by *Billboard* signaled the biggest change to sales counting methodology in the music industry since 1991. Like the changes in 1991, those in 2014 had serious implications for the music industry in general and rap music specifically. The *Billboard* charts and the business of pop music were revolutionized by the introduction of a new technology called SoundScan in June 1991. Before SoundScan, the *Billboard* charts were based on casual surveys of retail stores and reports from store managers. This data was consistently informal and imprecise. Occasionally, the imprecision was the result of sloppiness. In other instances the imprecision was a result of corrupt record execs paying store managers to inflate the sales of their company's records. SoundScan established a more formal and precise technology for tracking point-of-sale in retail stores. Every time a CD was scanned at the cash register, the point-of-sale transaction was counted.

The two genres that benefited the most from SoundScan were rap and country music. It turns out that the informal methods used by the industry to track sales were underreporting the sales of rap and country albums. Sound-Scan provided indisputable data that rap music sales were much bigger and whiter than anyone fully realized before 1991. This revelation changed the course of pop music history by recognizing rap as a mainstream form of entertainment.

The more recent decision by Nielsen and *Billboard* to track streaming music is having a similar effect in the music business. Two years after *Billboard*'s move to tracking multiplatform music consumption, the evidence shows how decisive a shift has been made to streaming music over other forms of consumption. According to the Record Industry Association of America, revenues from recorded music in the US increased 16.5 percent between 2016 and 2017 to a retail value of $8.7 billion. The driver of revenue was paid music subscriptions—more specifically, the rise of music streaming activity. In 2017 paid subscriptions to services like Spotify, Apple Music, Tidal, and Pandora increased by more than 50 percent from the previous year. Streaming music platforms accounted for 65 percent of total US music industry revenues in 2017. By contrast, physical sales (i.e., CDs, vinyl) accounted for 17 percent and digital downloads 15 percent.

No genre of music is streamed more than rap. Take a look at the major streaming platforms, and rap songs generally infiltrate the top of the charts. The status of SoundCloud rap, once an underground phenomenon, rose sharply as a result of the new industry metrics. The songs of several Sound-Cloud rappers, including Lil Pump and Lil Uzi Vert, have achieved impressive positions in the *Billboard* music charts. In 2017 Lil Pump's single "Gucci Gang" climbed up *Billboard*'s Hot 100 and Streaming Songs charts. Lil Uzi Vert's album *Luv Is Rage 2* topped the charts during its 2017 debut. In 2018 SoundCloud rapper XXXTentacion's album *?* debuted as the number-one album in the country. The overwhelming majority of his music sales were counted in streams, 61 percent, compared to album sales, 23 percent. Another SoundCloud rapper, 6ix9ine, debuted at number two on the *Billboard* album chart in 2018. Shortly after XXXTentacion was shot dead in June 2018, the *New York Times* reported that, according to Nielsen, his audio and videos had surpassed 4.6 billion streams, making him, posthumously, a bona fide superstar.

The success of SoundCloud rap precipitated what *Billboard* called a "gold rush" to sign underground rappers to label deals. In addition, the success of SoundCloud rap has spawned an entire subindustry made up of boutique labels, business managers, tour bookers, and publicists looking to cash in on the music's appeal. Virtually all of the chart topping and commercial success that the SoundCloud rap style enjoys can be attributed to a vibrant streaming ecosystem rather than physical sales and radio airplay, once the conventional path up the pop music charts.

The rise of SoundCloud rap signaled that a new era in pop music had arrived. In an earlier era, music industry executives pursued "spins"—radio airplay—with ferocious intensity. Spins were the marker of popular appeal and a predictor of music sales. But in the new world of digital music, streams are displacing spins as the barometer of popularity and commercial success. Rap music's commercial reputation has benefited enormously from the rise of streaming as the dominant commercial application in the music industry.

BACK TO THE FUTURE

In a bid to become a much bigger player in a rapidly growing digital music industry, SoundCloud management launched SoundCloud Go in 2015. The service was an attempt to become a streaming music destination for consumers like Spotify. The early adopters of SoundCloud—music creators—immediately called out management for abandoning the community that, in their view, built the platform. SoundCloud Go generated very little traction, which pushed the company to the brink of total collapse. In January 2017 the company announced that it was close to running out of cash. It turns out that users like SoundCloud rappers had figured out how to make the audio-sharing service work even as the company struggled to craft a sustainable business model.

The strategic missteps and insufficient revenue forced the company to lay off 40 percent of its workforce and close its San Francisco and London offices. SoundCloud management faced a dilemma: either run the risk of going under or devise a strategic overhaul of the company. The arrival of a new CEO and a cash infusion of $170 million in 2017 gave the audio-sharing service a second life.

In the aftermath of the shakeup, SoundCloud announced that it was recommitting to its original mission. More specifically, that meant building a product that empowers the creators of audio. The decision to embrace its original mission was heavily influenced by an unlikely group of young creatives—SoundCloud rappers—who were living proof that the platform could, in fact, launch careers in the music industry.

Some observers applauded the strategic return to the company's roots. Kerry Trainor, the new SoundCloud CEO, told London's *Financial Times* that "artists were always SoundCloud's core value, and that is how we need to measure its success." The kind of metrics that count, he added, include "how many people are starting their careers on SoundCloud?" Management decided that for the company to become viable in a rapidly evolving pop music economy, it would need to take its strategic cues from a group of upstart hip-hop artists who figured out how to turn the internet into an unlikely launchpad for a journey from total obscurity to micro-celebrity.

CHAPTER 5

THE PEOPLE'S CHANNEL

How an Awkward Black Girl Used YouTube to Prototype the Future of Television

Three PayPal employees started working on a video-sharing website on February 14, 2005. According to one of the founders of the site, the starting date was not that unusual for him and his two colleagues. "That's one of those things about being a computer science major: Valentine's Day, it's just another day, so why not start a new website?" he told graduates at the 2007 commencement for the University of Illinois Urbana-Champaign. Seventy-eight days later, April 23, the site went live when one of the founders posted the channel's first video, titled "Me at the Zoo." The clip was nineteen seconds long. At the time, the founders rented a single web server for $100 a month to host the site they called YouTube.

Steve Chen, Chad Hurley, and Jawed Karim imagined a site that would allow friends and family to casually share video clips online the way that millions of people were beginning to share photos online. They were confident that YouTube was a good idea, but they struggled to articulate exactly how video sharing might best be used. Their initial idea was to launch YouTube as a dating site. By 2005 the web was changing from something that people could only consume to a place where they could create. The proliferation of tools and platforms that enabled consumers of media to become producers of media led to whole new practices around the creation, distribution, and consumption of content. Media broadcasting, historically a top-down enterprise, suddenly developed a vibrant, bottom-up ethos. YouTube's initial tagline, "Broadcast Yourself," cleverly captured a cultural transformation that continues to evolve as content creators push the boundaries of what is possible in a rapidly evolving media landscape.

If the founders of YouTube struggled with the vision for their newfound site, the early adopters of YouTube did not. Users posted an assortment of videos, including clips of their pets, their vacations, spoofs of all kinds, and home video. When users began posting clips from popular television shows and television commercials, YouTube's role as an arbiter of pop culture tastes and trends was firmly established. The founders learned an important lesson early on in their execution, after trying to impose a top-down definition of YouTube. A few months after launch, the founders revamped the site and made it a more open, user-driven platform. It was at that point, Karim told students in his University of Illinois commencement speech, "Our little website was finally taking off."

THE PEOPLE'S CHANNEL

As the population of YouTube users expanded, the very idea of the channel— what it was and what it could be used for—also expanded. An early defining moment for YouTube was the creation of a web series by a teenage girl named Bree titled *Lonelygirl15*. From the privacy of her bedroom, Bree openly shared her thoughts about typical teenage topics, including her parents, school, and romantic interests. *Lonelygirl15* became a viral sensation, attracting millions of views. When a community of audience members grew suspicious of the authenticity of the videos, they started their own probe. Fans of the show eventually revealed that *Lonelygirl15* was a scripted hoax by three aspiring filmmakers looking to break into Hollywood. The hoax notwithstanding, the popularity of the series signaled, among other things, that creative storytellers would use YouTube in ways that the founders never could have imagined. Less than a year after it was founded, YouTube became a household name, a pop culture phenomenon, and a harbinger of the digital future.

On October 9, 2006, it was announced that Google had purchased YouTube for $1.6 billion, a staggering figure back then for a company whose financial future was still uncertain. Today YouTube is an iconic brand in a rapidly evolving global media landscape. It is currently valued between $80 billion and $100 billion. Each day, about one-third of all people on the internet watch a billion hours of YouTube and generate billions of views. Only Facebook draws more traffic globally than YouTube. In the US the video platform reaches more eighteen- to thirty-four-year-olds—a highly coveted and influential demographic—than any cable network.

YouTube has also become what I call "the people's channel"—that is, a multidimensional platform broadcasting all manner of content—short-form stories, music, indie docs, fashion and fitness vlogs, games, Hollywood trailers, science—from all over the world. Roughly four hundred hours of content is uploaded to YouTube . . . every minute.

The world's biggest online video channel is also a central node in the bustling ecosystem that fuels the new innovation economy. Successive waves of innovators have adopted YouTube to reimagine everything from media and entertainment to education and politics. A cast of new and unique voices leverages the power, appeal, and scale of online video to inform, incite, entertain, and empower communities worldwide. In 2011 one of those content creators was a young woman living in Los Angeles who had dreams of telling stories from the perspective of an awkward Black girl.

She had pitched to television executives before, only to be told that there was no audience for such a program. A few years earlier, before the rise of social media and the do-it-yourself ethos they inspire, that kind of rejection might have meant the end of her dream. But instead of giving up hope, Issa Rae turned YouTube into her own innovation lab, illustrating how the channel once welcomed sensibilities and stories that sharply contrast the material so often found on mainstream broadcast channels.

BLACK AND AWKWARD

Issa Rae's personal story is a familiar one among millennials. She graduated from Stanford in 2007 just as the Great Recession was rendering many young workers—those with college degrees and those without—vulnerable in a contracting labor market. Like a growing number of college graduates, Rae experienced chronic underemployment. The jobs that she found did not require the degree that she had gone into debt to earn. "I was just working from job to job, trying to find money where I could. And freelancing a lot," Rae told the *Huffington Post* in 2013.

Soon after graduation Rae moved to New York to work for a nonprofit theater company. She aspired to do more. A goal of hers in New York was to turn a web series she created while at Stanford called *Dorm Diaries* into a TV program for cable. Not surprisingly, making it in New York was daunting. She struggled to get executives to take her seriously. One day someone broke into her apartment and stole her laptop. For a young creative like Rae this was tantamount to taking her studio. All of her videos, demo reels, scripts, and ideas were on the laptop.

New York, in her words, "wasn't going like I really hoped." Struggling professionally and unsure what to do next, Rae found herself in her shoebox-size apartment writing in her journal when she penned the words that would eventually change the course of her life. "I'm Black. And I'm awkward." Rae would later describe the post to her journal as an epiphany, the moment she realized who she was personally and creatively. Reflecting back on that moment, Rae joked, "I knew I was black for a long time, but I realized I was awkward just in the middle of the day. I was sitting in my bed, in my small closet-apartment in New York, and it just came to me." The "awkward and Black" theme was an idea around which she would eventually kickstart her own mini-creative economy—merchandise, a web series, campus lecture tour, a best-selling book, podcasts, and social media influence. Five years later the culmination was an original series for HBO titled *Insecure*.

A seed was planted that day in her apartment, but the realities of life—economic survival—delayed developing the "Black and awkward" theme into something tangible. After all, she was super broke and struggling to hold a steady job. Sensing that New York was not going to work out, Rae moved back to Los Angeles. If she was going to fail, at least she had a safety net of family and friends on the West Coast.

In Los Angeles, Rae worked as a temp. But she continued to make media. She started another web series that chronicled her brothers' hip-hop band and their struggle to make it in the Los Angeles music scene. "I was really struggling. Super, super broke and trying to hold onto a job," Rae recalled a few years later.

One day on Facebook a friend shared a link to an article from *TheGrio* magazine with Rae. It turned out to be the most important social media post of Rae's life. The article asked, "Where's the Black Version of Liz Lemon?" a reference to the likable nerd played by Tina Fey on the television show *30 Rock*. The journalist who had written the article, Leslie Pitterson, challenged Hollywood to invest in the Black nerdy girl. Pitterson added, "I think my little sister, and all the other less than put together brown girls growing up now, deserve a heroine who shows them the awkward journey is not theirs alone." Pitterson's article was a sharp critique of the narrow representation of Black women in pop media culture.

Rae panicked. This was her idea, the one she had come up with two years ago in her New York City apartment. "What if someone else read this and thought the same thing?" she asked herself. If she was ever going to turn the "Black and awkward" theme into something tangible, it was now or never.

For two years Rae had been stymied by what she did not have—money, a production team, time, or a studio deal. After reading Pitterson's article, she shifted her attention to what she did have. As she told a group of young aspiring content creators in 2014, "I owned a camera, I owned a computer, I owned editing software, I owned the internet, I owned friends, I owned twenty-five dollars to buy them lunch."

And that's what she used to produce the first episode of *The Misadventures of Awkward Black Girl.*

YOUNG CREATIVES AND THE NETWORK EFFECT

Rae knew instantly that she would distribute *The Misadventures of Awkward Black Girl* through the web. "I tried to pitch one of my other web series to TV, and I was met with certain ideas and certain forms of criticism that I didn't necessarily agree with," Rae told a reporter from *Time* magazine. Hollywood, from the perspective of young underrepresented talent like Rae, was a barrier to opportunity rather than a creator of opportunity. "The internet," she would say later, "is where you can find what you're not seeing in TV and film."

In order to make her web series a reality, Rae did what many young aspiring entrepreneurs do when they think they have a great idea but no money: she turned to her network of friends and acquaintances. This is what I call the "real sharing economy." Financial capital is certainly crucial in the new innovation economy, but it is also elusive. There are no angel or venture capital investors available to young creatives on the come up. As a result, social capital—the resources one is able to extract from the people they are connected to—becomes a vital asset. In the midst of great economic uncertainty, young creatives share their time, talent, knowledge, and labor to help friends or acquaintances pursue their creative or entrepreneurial ambitions.

Rae's story is an excellent example of how young creatives maintain social ties that are rich and diverse, despite claims that they are tech addicted and socially isolated. More specifically, Rae relied on two crucial sources of social capital: her personal network and a more distributed network made possible via the internet.

Her personal network included strong ties (friends and family) and weak ties (e.g., friends of friends) that she was able to rely on for support. This closer network of friends and acquaintances provided the material labor to make *Awkward Black Girl* a reality. For example, Rae called on one of her best friends from high school to shoot the first episode of *Awkward Black Girl* guerrilla style, a form of filmmaking characterized by low budgets, makeshift crews, few props, and whatever resources are available. She also reached out to a friend from her Stanford days to play a key character in the web series. Rae had studied film at Stanford and even produced a couple of web series already, but she was still a novice. The production team that she cobbled together consisted of amateurs who were all young, hungry, and ready to make a move in life. When the pace of production eventually ramped up, Rae relied on her close social ties to help provide the writing, camera work, acting, equipment, and emotional support that she needed to produce *Awkward Black Girl*. A few of them even made small financial donations.

Rae also relied on a more distributed network of people, the kind that only the internet can enable. This source of social capital generated the audience, community, and financial backers that made *Awkward Black Girl* a reality. This network, made up largely of people Rae did not know and would never meet, provided vital resources too. Her internet fan base showed massive support for the series via their views and YouTube comments. Later, when Rae needed money to sustain the web series, it was the crowd power made possible by the internet that provided financial support.

PROTOTYPING THE FUTURE OF TELEVISION

As the audience for the web series grew, so did the creative vision and confidence of Rae and her coproducer, Tracy Oliver. Their growing faith in the project was evident, for example, in the increased episode length, richer story lines, and more professional production values. The writing sharpened, and the camera work, lighting, editing, and sound quality also improved. As the producers approached the end of season one, they began to think less in terms of a sketch series and more in terms of an episodic structure. In short, they began to envision *Awkward Black Girl* as a television show, but one available on a connected device.

The lack of a budget forced Rae to continue to shoot the web series guerrilla style, which had some notable advantages. For example, shooting in small apartments and office spaces imbued the web series with a sense of intimacy and authenticity. The audience responded instantly, expressing approval for the humor and attention to the small ways race, gender, and class distinctions matter in the lives of millennials. They liked the show's whimsical characters. Rae plays J, whose chocolate skin tone, short-cropped Afro, and awkward style make her an unlikely lead character. The white office manager's efforts to connect with her mostly young and Black and brown workforce are typically marred by unintentional but definite racial slights. J's coworker Nina, played by Oliver, is the villain. Her light complexion, straight hair, and sorority background smack of Black elitism.

Among fans, there was also an appreciation for *Awkward Black Girl*'s cultural sensibilities, including inside jokes, razor-sharp reflections on the subtle aspects of race in everyday life, and the hip-hop vibe that added texture to the show. *Awkward Black Girl*'s narrative machinations, rainbow-colored cast, and quirky disposition were a refreshing contrast to the standard television fare. The writers probed the micro-details of millennial life through the perspective of a young Black woman. Along the way, the web series adroitly engaged the theme of chronic underemployment that has come to define the millennial experience. In this case, the series shows college-educated people working in soulless jobs that offer little opportunity for economic mobility. J works for Gutbusters, a company that manufactures diet pills. When she and a colleague are teamed together to produce a new marketing campaign for the company's new colon-cleanse pill, they develop the slogan "I Got My Colon Clinzed!!" Though satirical in nature, jokes like this one highlight the mind-numbing work many millennials face in today's "knowledge economy."

As the narrative arc of the web series expanded, issues related to interracial dating, race and beauty aesthetics, and workplace politics emerged. But *Awkward Black Girl*'s narrative voice was never didactic. It did not need to be. Seeing young Black, East Asian, and white characters laboring in tight cubicles, living in modest apartments, and navigating the challenges of professional and personal relationships was a shrewd counter to the limited and token representations that characterized mainstream scripted entertainment.

Rae's play with rap music is another notable feature of the web series. She frequently uses rap as a source of comic relief but in ways that also illuminate the music's vitality in the lives, most notably, of young Black women. More specifically, the clever use of rap music conveys a bold Black female sexuality that at that time was virtually unrepresented in Hollywood. Rae's character J listens to rap music but also writes her own rhymes as a form of therapy, anger management, and sexual desire. These and other factors illustrate how the web series references a specific culture—hip hop—while also addressing universal themes—female empowerment.

Posting episodes on the web established an opportunity for Rae to produce a working prototype of her vision for what television should look like:

more diverse and culturally incisive. Importantly, the web also provided an opportunity for Rae to simultaneously distribute *Awkward Black Girl* and iterate the quality of the production with the help of constant audience feedback, which was immediate, honest, and influential. For example, when audience members responded favorably to the introduction of White Jay (one of Rae's character J's romantic interests), the writers expanded his role. Audience feedback also commented on the poor audio quality in the first few episodes, compelling Rae's team to be more attentive to the web series production values.

Through the web series, Rae became a vibrant symbol for young Black women, who found her portrayal of an awkward Black girl refreshing, relevant, and a provocation for laughter and contemplation. The social media community and conversation that *Awkward Black Girl* sparked was tantamount to a series of real-time focus groups marked heavily by Black female frustrations and aspirations. Many young Black women saw themselves reflected in the web series in ways that simply were not possible in mainstream entertainment. Among the community that rallied around the web series via Rae's Kickstarter campaign, one woman described *Awkward Black Girl* this way: "It was like watching something about me and people I know. Totally real and completely awkward. Thanks for taking the time to create such an honest, hilarious, and much needed series." A Black female graduate from Stanford also found value in the web series, noting, "I appreciate what your show means for every black [woman] out there who felt invisible because they never fit the stereotype of what black women are 'supposed' to be. Hopefully there will be many more episodes of ABG for all of us!!"

SERIOUS ABOUT SOCIAL MEDIA

Like many digital-content creators, Rae had to become a discerning student in the art and science of storytelling via the internet. The science dimension of the project included figuring out ways to grow the number of people who tuned in to watch the web series. The art dimension included figuring out ways to connect to and foster the community that would advocate for the web series in their social networks. Rae told *Fast Company* in 2012, "People aren't going to watch a video that takes 7 minutes if they don't know anything about it. It's much better to start with something only a few minutes long." She added, "I learn something new every single day when I'm doing this, honestly. Because, just, nobody has mastered the art of it."

What she was learning—how to use social media—is something that young creatives have been studying informally since the rise of the medium in the early 2000s. Social media is full of data to be mined, understood, and utilized. When the web first took off, the metrics were relatively straightforward. Did someone click on your site? How many pages did they view? How many unique visitors did you attract? But as the population of social media users has expanded, so have the forms of social media engagement. As a result, the methods for understanding social media behavior have grown more sophisticated, requiring nuance and context-specific understanding in the era of "big data" and "social data." Audience-generated content—"Likes," comments, tweets, retweets, hashtags, memes, GIFs, photos, and videos— generates enormous opportunity for creatives to gain insight related to both the quantity and the quality of user engagement. This is certainly what young creatives like Rae have learned as a result of their investment in social media. Big tech companies like Facebook have built entire units and their fortunes from their efforts to better understand social media engagement.

When Rae began to think about producing a web series, her attitude about social media became, in her words, "more serious." Like many other young social media enthusiasts, she began to develop a richer and more expansive view of social media. Rather than seeing it as something primarily for leisure use, she recognized that it was a resource for pursuing her creative and entrepreneurial aspirations.

The most tech-savvy young creatives are not simply social media enthusiasts; they are also social media analysts. They have learned to be attentive to the nuances of influence in the social media economy. The ability to detect key patterns in social media content, audience, and feedback has become a form of cultural literacy that pays in cultural currency. Cultural literacy refers to the knowledge and expertise that one gains as a result of using social media, while cultural currency refers to the strategic ways in which that knowledge and expertise are put to use.

Rae and her team learned quickly that social media was a resource for building, engaging, and learning from their audience. Even after Rae graduated to HBO, she and her staff continued to take social media seriously and even monitored the audience's live tweeting behavior during the airing of *Insecure*. Live tweeting functions as a real-time focus group, allowing the show's writers and producers to gauge which story lines, character dialogue, jokes, and even musical choices resonate with viewers.

This is a whole new skill set that young creatives have developed to translate social data—first into insights and then into creative action. This knowledge functions as a vital form of expertise as social media assumes an increasingly central role in communication and everyday life. Today, most organizations hire social media managers—an occupation that young creatives have played a significant role in making relevant in the digital economy.

CHANGING THE FACE OF TELEVISION

Initially *Awkward Black Girl* was Rae's side hustle. She worked on the series late at night, on the weekends, and whenever else she could find the time. Much to her surprise, *Awkward Black Girl* became a YouTube sensation. As the audience for the web series grew, so did the demands on Rae's time. She quit her temp job. "It's so hard to have a nine to five and try to be creative in the evening, or schedule everything on the weekends because you have to," she said.

Before she knew it, Rae's side hustle had become a full-time gig. But with no budget to speak of, *Awkward Black Girl* became, in reality, a full-time struggle. This was before YouTube had developed the infrastructure to support and cultivate the people known as "YouTube stars," a growing cast of individuals who have turned the platform into a launching pad for content creation, branding, and celebrity.

The success of the web series was a blessing. Rae had touched a nerve among an audience of viewers who identified with her notion of being Black and awkward or just simply appreciated a voice and viewpoint about contemporary life that could only be seen on the internet. But the success of the web series was a curse too. The growing audience required her modest production staff to produce episodes at a more regular pace to sustain the momentum. No one was paid for his or her time, talent, or service. *Awkward Black Girl* was clearly entertaining, but the production quality—sound, camera, editing—reflected Rae's humble circumstances. Viewer growth demanded an upgrade in the series' production values. Demands like these required the one thing that Rae did not have—money.

After finishing post-production for season one, episode six, Rae and her coproducer came to a crossroads. They were both jobless and could no longer continue funding *Awkward Black Girl* out of pocket. It was at this point in the process that they turned to crowdfunding. Among the many interesting developments powering the new innovation economy, none may be more representative of technology's influence than crowdfunding. To make her crowdfunding campaign distinct, Rae framed her bid for support on Kickstarter this way:

> Television today has a very limited scope and range in its depictions of people of color. As a black woman, I don't identify with and relate to most of the non-black characters I see on TV, much less characters of my own race. When I flip through the channels, it's disheartening. I don't see myself or women like me being represented.

Rae's Kickstarter statement addressed a deep dissatisfaction with television representations of diversity. In a brilliant move, she framed her crowdfunding campaign for *Awkward Black Girl* as a call to action. Rae's crowdfunding campaign was more than a money-raising endeavor; it was also a political endeavor. In the campaign video she called on her backers to "help us change the face of Hollywood."

The strategy was likely informed by the analytics Rae and her team generated from the popular following they amassed via social media channels like YouTube, Facebook, and Twitter. Many of the comments that circulated through social media echoed a common sentiment among *Awkward Black Girl* audiences: they liked the show because it portrayed diversity in a humorous, creative, and relevant way. The "Black and awkward" theme mobilized a substantial and passionate community for the web series that became a powerful source of social and financial capital. More specifically, the *Awkward Black Girl* community deployed the power and scale of the internet to spread the word about the series. In addition, the community rallied around the crowdfunding campaign that ultimately sustained production for the web series.

The profile of the community of backers illuminates the diversity and sense of urgency that powered the campaign beyond its initial funding goal of $30,000. A total of 1,960 people backed the *Awkward Black Girl* Kickstarter campaign and raised $56,259. A decisive majority of them, 82 percent, were first-time backers, suggesting that *Awkward Black Girl* tapped a community that was new to the world of crowdfunding. The largest concentration of backers came from major metropolitan centers such as New York, Los Angeles, Brooklyn, Washington, DC, Chicago, and Atlanta. These cities have two main things in common. First, they are hubs for millennials, especially the aspirational and creative types. Second, cities like Washington

DC, Chicago, and Atlanta also contain large concentrations of young African Americans, an early and pivotal audience for *Awkward Black Girl*. Almost half, 46 percent, of the backers were from countries outside the US, including Canada, Great Britain, South Korea, and Australia, a testament to the global reach of the internet.

The donations by the backers of *Awkward Black Girl* amounted to a vote for a storyworld they knew Hollywood would never support of its own volition. As a result, backing *Awkward Black Girl* became tantamount to joining a cause to not only change the face of Hollywood but, in this specific instance, also to change the face of television.

THE REAL DISRUPTION

Rae's celebrity status is based, first, on *Awkward Black Girl* and then on becoming the first African American woman to have an original series broadcast on the premium cable network HBO. But her potential for deeper influence is linked to an endeavor that is largely beyond public view. It is a project Rae calls Color Creative. The initiative is Rae's bold attempt to leverage her pop culture capital and harness the power of the internet to build an entertainment ecosystem that cultivates underrepresented creative talent, such as writers and directors. The most disruptive innovation among young creatives like Rae is how they leverage social media to build systems that support diversity and opportunity. Like other young creatives in different spheres, Rae's innovative impulse is a disruptive one. Her goal with Color Creative is to shake up an industry that remains notoriously white and exclusive and prove the wide appeal of greater diversity.

"I started Color Creative to give opportunities to talented women and writers of color," Rae told *Fast Company*. "I get tired of hearing that we don't appeal to a 'broad audience,' whatever that means. We're providing opportunities and showcasing stories that aren't being told anywhere else." Rae came up with the idea for Color Creative in 2013. Some of the scripts that she read had potential but no credible path to studios or audiences. She could identify with the writers, because she had been one of them. She envisioned Color Creative as an "independent television network disrupting traditional models to make them more inclusive." Color Creative might be best described as an incubator for underrepresented creative talent.

Color Creative is a potentially groundbreaking project because it uses an internet platform to diversify the talent pipeline for one of the most homogeneous occupations in Hollywood, television writers. According to a 2017 study authored by UCLA's Darnell Hunt, Black writers are basically absent in Hollywood. The study, titled *Race in the Writers' Room*, examined 1,678 first-run episodes for all 234 of the original scripted comedy and drama series airing or streaming on eighteen broadcast, cable, and digital platforms during the 2016–17 television season. The study reports that two-thirds of all shows had no Black writers in the writing room. Among all of the writers on all shows, 3,817 in total, only 14 percent were people of color. Blacks made up only 5 percent of all writers.

The writer's room is not only where stories are created, it is, for all practical purposes, where the fictional world of pop culture is made. The lack of diversity in the writer's room explains a lot about the lack of diversity in the stories and sensibilities that shape our entertainment landscape.

One of the main challenges that aspiring writers and directors face is an inadequate social infrastructure—lack of connection to industry professionals, mentors, and sponsors. These industry brokers guard access to the knowledge, expertise, and professional experience that drive opportunity for those on the outside. While the talent and good ideas that make up human capital certainly matter for aspiring young creatives in the entertainment field, personal connections—social capital—also matter. Referring to her own early struggles, Rae says, "I was trying to break into the industry traditionally by writing and entering contests, and quickly figured it is really a 'who you know' industry." As she highlights here, aspiring writers, for example, pay a high cost for not knowing the right people. To really impact the plight of underrepresented talent, innovations like Color Creative must not only develop creative potential or human capital; they must also develop industry connections or social capital.

The core principles of Color Creative—tech savvy, collaboration, inventiveness, disruptiveness, and commitment to diversity and inclusion—reflect the ethos of the new innovation economy. Color Creative is an interesting concept because it is designed to leverage the power and scale of the web to create something that underrepresented creative talent in media entertainment has never really had—an ecosystem that fosters the cultivation of human and social capital. In order to have substantive impact, Color Creative must be an incubator for sharing ideas, modeling successful projects, providing genuine mentorship, sponsoring untapped talent, and connecting them to real opportunity. In short, it must be an engine for creativity, connectivity, and opportunity.

Ideally, it would do all of this at scale.

Rae was not the only young creative to use YouTube to invent a distinct vision of media storytelling and entertainment. But she was making *Awkward Black Girl* during the period before YouTube strengthened its commitment to corporatization and algorithms engineered for monetization. Today YouTube is striving to become a major player in the corporate media economy. Once a disruptive start-up, it has repositioned itself as a major broadcast player. For example, through the YouTube Partner Program (YPP) and YouTube TV, the company's management has redirected the platform away from upstart media makers and toward legacy TV production and a roster of YouTube stars whose projects comply with the company's recent stricter requirements. In January 2018 YouTube announced these new eligibility requirements: in order to make money through ads and the YPP, a content creator must have amassed four thousand hours of watchtime within the past year and a thousand subscribers. Critics charge that YouTube has redesigned the platform to support and reward content providers who are risk-averse and thus less likely to offend advertisers. This is, in effect, the old broadcast model. As a result, a new creative hierarchy has emerged at YouTube; the corporate structure now invests its resources—its maker studios, ad revenue, and algorithmic-driven recommendations—in culturally safe content creators. These developments tend to shut out upstart content creators like Issa Rae who push the creative envelope to expand our notions of entertainment.

CHAPTER 6

CAN YOU HEAR US NOW?

How Crowd Power Shatters Hollywood's Perception of Black Audiences

Netflix is at the vanguard of a new era in television. With its use of streaming, the binge viewing model, and its algorithmic recommendation engine, Netflix has transformed how we watch television. Unlike its competitors in broadcast and cable, Netflix is not beholden to the traditional advertising model and the restrictive sensibilities that limit the kinds of content we see on broadcast stations. On April 28, 2017, Netflix began streaming *Dear White People*, a new original series. The series was based on the film with the same title that was written and directed by Justin Simien three years earlier. Simien was named one of the executive producers and was also contracted to direct three of the ten episodes. *Dear White People* is a satirical take on the peculiar state of race in millennial America. Among other things, the film follows the experiences of four Black students—each distinct in terms of their racial identity and the paths they take to navigate life at a majority-white elite college.

Netflix's decision to stream original episodes of *Dear White People* did not go over well with some white people. Fresh off a 2016 presidential election that awakened white nationalists, there was a modest attempt among some from the alt-right to make Netflix pay for *Dear White People*. For instance, there were calls to cancel subscriptions, but the movement did not sustain any real momentum or threaten Netflix in a meaningful way. Unwittingly, the feeble protest may have served to embolden Netflix to embrace more diverse talent like Simien.

Netflix asserts great influence in the media entertainment landscape. But even as the entertainment streaming service changes the way we watch television, one thing remains stubbornly constant: Netflix's lack of diversity in the creative talent it recruits and in its management ranks.

A 2017 report by the Directors Guild of America ranked Netflix last among the ten largest television studios in terms of hiring Black and Latino directors. White men directed 69 percent of the episodes of original programming on Netflix in 2015 and 2016. By comparison, persons of color directed only 14 percent of the episodes. Women, too, made up only 14 percent of the talent who directed Netflix episodes. Netflix has also been criticized for the lack of diversity and inclusion in its management ranks. African Americans and Latinos make up 4 and 6 percent, respectively, of Netflix's staff and leadership team. *Bloomberg* reported that just "one of the all male executives taking part in quarterly earning calls with investors is nonwhite." In 2018 the streaming giant fired its chief communications officer after it was revealed that he used racial epithets on more than one occasion during company meetings.

When it comes to adequately addressing lack of diversity and inclusion, the future at Netflix is unclear. But the entertainment model that Netflix pioneered, streaming and data-driven entertainment, has proven its worth. In fact, the company is a victim of its own success. Look no further than the growing list of competitors entering the streaming space, which includes legacy broadcasters like Disney and upstart content providers like Apple and Facebook. Recognizing the rise in competition, Netflix announced that it would be spending more than $8 billion on original programming in 2018, a stunning amount. But one of the questions that the company is certain to face is to what degree it invests in young and diverse showrunners, writers, and directors. To what extent will the management that the company invests in reflect diversity and inclusion?

The new innovation economy built by young creatives is compelling companies like Netflix to be more responsive to the currents of cultural and political change. Simien's presence on Netflix is a good example. *Dear White People* is certainly a product of Simien's tech and social ingenuity. But it has also been influenced by something I call "Black crowd power," more specifically, the growing influence of Black online communities, what scholars call "networked publics," in the realms of pop culture and politics.

Simien's road to Netflix was paved six years earlier when he set out to make a film that, in his words, "no one was looking for." Similar to Issa

Rae's experience, Simien's struggle to make the film highlights some of the more compelling characteristics of the new innovation economy, such as the side hustle, the inventive use of social media, and the desire among young creatives to hack the business-as-usual ethos that shapes many of the key institutions in America, including Hollywood.

FEEDING THE CREATIVE BEAST

Simien's dream of becoming a filmmaker began during childhood. "I remember I was watching TV and all of a sudden it occurred to me that it was someone's job to actually put the stuff I was watching onto TV. . . . When I realized that movies and TV shows were an actual thing that people did as a job in life it blew my mind. It's all I've ever wanted to do since." His passion for visual storytelling continued throughout high school and when he attended Chapman University to study film. Located in pristine Orange County, Chapman is just a few miles south of Los Angeles, the epicenter of the film and television universe.

Simien's experience at Chapman was formative. Coming from an arts magnet school in Houston in which his friends looked like "a Benetton ad," the "white as snow" setting at Chapman was a culture shock. Black students made up a tiny fraction of the total student body. "It wasn't even that it was white, it was people who had not had a lot of interaction with people of other cultures, whatever that culture may be," Simien told *Interview*.

Life at Chapman was a constant source of conversation for Simien and his African American peers. The environment was not necessarily hostile, but it was characterized by what sociologists call micro-aggressive behaviors. These are the subtle but common verbal and nonverbal slights and insults that reflect hostile or indifferent attitudes toward those perceived as inferior or subordinate. At Chapman Simien found that the micro-aggressions of whites in their interactions with Blacks were often the result of the lack of exposure to and familiarity with racial, ethnic, and cultural difference. A white person asking a Black person, "Can I touch your hair?" would be an example of micro-aggressive behavior. A white person assuming that a Black student was admitted to the university because of affirmative action is another example. In both cases Blacks are treated as different from and inferior to their white peers.

In one of his film courses Simien began to explore what it was like to be a "black face in a white place" in a script that he titled *2%*. The title was a reference to the low number of Black students on elite college campuses across the US. Over the next few years as he developed the script, Simien began doing research on the experiences of Black students attending Ivy League schools. Even though Black students were present on all of the elite campuses across the US, they were still marginalized and subjected to casual and institutional racism.

During his research Simien discovered the high rate of blackface parties held by white students. These parties were a modern-day expression of what historian Eric Lott calls the "love and theft" legacy of blackface minstrelsy in American culture. Lott argues that minstrelsy is a peculiar phenomenon marked by fear of the Black body on the one hand and fascination on the other. The blackface parties on college campuses often exploit the popularity of hip hop, especially the exaggerated posturing associated with gangsta culture and "thug life," while simultaneously mocking the conditions of Black urban poverty. In a later version of Simien's script, a Black-themed party at a fictional Ivy League college would be a pivotal source of narrative conflict.

After graduating college in 2008, Simien found steady employment laboring away in the publicity departments of studios like Focus, Sony, and Disney. His work in the bowels of Hollywood was not soul crushing, but it was not soul lifting either. Simien desperately wanted to make a career pivot. He wanted more. He wanted to be a storyteller. Simien told himself, "I need to do this. I can't spend another year wishing I had done this. And even if it takes me ten, twelve, a hundred years to do this, that next year is going to be so much better than it otherwise would have been."

After graduating college Simien had stayed active creatively. While working his paying gig, he wrote small screenplays, produced short web series, and edited the film shorts of friends. Pursuing his creative instincts became a ritual. Reflecting on this period between college and the creative success he dreamed of, Simien told a group of aspiring filmmakers, "Whether the script gets sold, or the pitch gets picked up, find a way to enjoy sitting down in front of the computer to write, the iPad to read, or whatever it may be." Simien added, "Talent and tenacity are important, but consistency of effort is what takes a project to the finish line." Simien has a term for this effort. He calls it "feeding the creative beast."

After nearly eight years of working in publicity, Simien picked up the script that he had written as a student at Chapman. The next two years would require an enormous amount of grit to turn a fledgling college screenplay about race in millennial America into a feature-length film.

REJECTING THE POSTRACIAL NARRATIVE

Before Simien knew it, turning his old college script into a project that he could pitch to studios became his side hustle. He labored on the script whenever he could find time, which was usually after work and late at night. He gave up weekends to do research and write. Simien's goal was aspirational and humble, he later recalled: "To get the film done and be able to pay my rent."

The story that Simien sat down to write was a satirical take on the state of race in so-called postracial America. Set at Winchester University, a fictional Ivy League school, *Dear White People* shines a spotlight on white racial privilege and micro-aggressive behavior while also exploring the many dimensions of Black racial identity. The film's lead character, Samantha, or Sam, is a biracial student who has been elected president of the Black student dorm. She takes the role seriously. Sam is also the creator of a controversial campus radio show titled *Dear White People*. On her radio show, Sam frequently challenges the racism expressed on campus and beyond, making her a target of white resentment and backlash. Privately, Sam dates a white student and maintains interests that do not necessarily reflect her more strident public persona as a Black student activist.

Sam's counterpart in the film is Coco, a dark-skinned African American who aspires to acceptance from her white peers and who wears blue contact lenses and a straight-hair weave. Simien's treatment of Sam and Coco provoked charges that he failed to deal adequately with the effects of colorism and privilege within the African American community. Some critics accused him of investing more time in the lighter-skinned character, Sam, to the neglect of her darker-skinned and equally compelling counterpart, Coco.

Simien's film about Black life in a college setting was not unprecedented. Other Black filmmakers have explored the complexities associated with being Black on a college campus. In 1974 Warrington Hudlin made a film titled *Black at Yale*. Hudlin and his brother, Reginald, would go on to ride the wave of what Simien calls Black smarthouse films of the late 1980s and early 1990s—movies with an art-house sensibility that were distributed in multiplexes. While Spike Lee's *School Daze* (1988) was situated on a Black campus rather than a majority-white campus, it, too, dealt with the various shades of Black identity, culture, and politics in the context of the college experience. Lee, a source of inspiration for Simien, is a cinematic provocateur, and his films acted as the main catalyst for the Black smarthouse films of that period. John Singleton, best known for *Boyz n the Hood* (1991), explores Black and white racial conflicts on a college campus in *Higher Learning* (1995). Building on that tradition, Simien's contemporary, Issa Rae, made a web series about Black student life at Stanford titled *Dorm Diaries*. Simien also drew inspiration from other film comedies set in school, including *Animal House* (1978), *Rushmore* (1998), and *Election* (1999).

If the thematic terrain of *Dear White People* had been previously traveled, Simien offered a fresh voice that reflected the complexities of race and identity in millennial America. *Dear White People* highlights the fact that despite the mainstreaming and commodification of Black popular culture—music, dance, fashion—and the election and reelection of the nation's first Black president, race continues to matter among millennials, the most racially and ethnically diverse generation in US history. More specifically, Simien's film rejects the postracial narrative.

After finishing the script, Simien had to confront the fact that his film was not likely to generate interest among studios. Simien believed there was an audience for *Dear White People*, but he knew that he needed to pursue an unconventional path if he was ever going to build that audience and get the film in theaters.

CREATING A TRAILER FOR A MOVIE THAT DOES NOT EXIST

Simien did not have the resources—human, social, or financial—to turn his script into a feature-length film. So he decided to do something outside the box: He took his 2011 tax returns, about $2,000, and used it to finance a concept trailer. He told the crew he had gathered, "Let's make a trailer treating the movie as if it already existed."

A concept trailer for a movie that does not exist is tantamount to what the tech and entrepreneurial world call "rapid prototyping," which allows innovators to make a version of their idea quickly and cheaply. It reflects the "fail fast" ethos of the tech and entrepreneurial world, the idea that in order to bring an innovation to life you should make it quickly, test it frequently, and constantly iterate. In the lead-up to taking Facebook public in 2010, Mark Zuckerberg called this philosophy the "hacker way." Zuckerberg wrote that this approach to innovation is about "building something quickly or testing the boundaries of what can be done." By prototyping his film with the concept trailer, Simien was testing the boundaries that restrict what stories get told in popular entertainment. The concept trailer, like all rapid prototypes, was intended to move the film forward by putting a small version of it in the world and eliciting feedback.

In addition to $2,000, Simien had some friends and some experience making short films. With a few crew members and some local actors, Simien went to the UCLA campus to shoot the trailer for a movie that existed only in his head. Campus police kicked them off campus because they did not have a permit, but the crew went back the next day to finish the shoot. The two-and-a-half-minute trailer artfully mimics clips from a variety of scenes in the script.

The finished product looked like a trailer for a film that actually existed. Simien said that when he made it, he did not want to create "the sort of backyard concept trailer that you see on YouTube, but like a real, like, this feels like it was cut from a real, finished movie." Just because the concept trailer was a do-it-yourself job did not mean it had to look amateurish.

Simien's next move was also typical in the new innovation economy. "Instead of using the concept trailer to just take meetings I said, 'Fuck the meetings. Let's put this directly on YouTube and see what happens,'" he explained. Simien elected to take the trailer to the people's channel to elicit feedback and gauge support. He believed that if he could get between 100,000 to 200,000 views, he could make a powerful data-driven case that there was a viable audience for *Dear White People*.

The response to the trailer would exceed anything he could have ever imagined.

WE'RE MORE THAN TYLER PERRY

Simien's decision to put the trailer on YouTube made sense. The world's biggest television platform is where millennials go to satisfy their interests in music, film, games, and television. It is also where they find content that connects with them in a more personal and sometimes more meaningful way.

Simien turned away from the gatekeepers and toward the crowd, that massive collection of people that has become, via the web, a prominent source of power and influence in the culture. Importantly, a significant portion of the crowd he tapped happened to be Black millennials. Much to Simien's surprise, after he posted it on YouTube, the trailer went viral and provided the momentum to launch a crowdfunding campaign to accelerate preproduction for the movie.

Simien kicked off his campaign with this provocation on the crowdfunding platform, Indiegogo: "Remember when Black movies didn't necessarily star a dude in a fat suit and a wig? Or have major plot twists timed to Gospel numbers for no apparent reason? No? Damn . . ."

This was a not-so-subtle reference to the dominance of Tyler Perry's movies, Hollywood's go-to but limited conception of Black entertainment. The industry's investment in Perry demonstrated a lack of appreciation for the complex sensibilities of Black millennials.

Similar to Issa Rae's, Simien's approach to crowdfunding was entrepreneurial and political. Like many other Black millennials, Simien had grown weary of the Black-themed films that Hollywood had embraced. He explained it this way: "Hollywood is a place where whatever worked last year they're just going to keep doing it, particularly when I started the film the only thing that was really happening was Tyler Perry in movies. And that's not really a dis to Tyler Perry, it's more that the industry didn't seem to have room for anything else."

Simien, like Rae, connected with a generation that had grown fatigued with Hollywood's limited notions of Black-themed entertainment. A survey of the comments from those who contributed to the *Dear White People* crowdfunding campaign suggests that the concept trailer connected with a desire for greater diversity and complexity in the representation of Black culture and life. The backers viewed funding the film as a strong rebuke of Hollywood's lack of diversity behind and in front of the camera. Among the community that mobilized to fund *Dear White People*, there was widespread discontent, for example, with Perry's reign as the arbiter of Black taste and scripted entertainment. Black millennials, quite simply, were eager for something different and relevant. The satirical reflections on race and society in *Dear White People*, along with a compelling narrative voice and style of humor, certainly met that standard.

The trailer was a tease, according to Simien: "Hey, this great movie is coming. Psyche . . . it's not actually coming but if you would like for it to come donate . . . a couple of bucks." His crowdfunding goal was $25,000 to complete some of the preproduction elements such as casting, retaining legal services, hiring a line producer, research, and executing a star-studded table read to attract financiers.

In its first day of official launch, the *Dear White People* crowdfunding campaign reached a quarter of its funding goal and generated more than 50,000 views on YouTube. Within three days the campaign generated more than a million views on YouTube and $25,000. Simien, awed by the fast response, thanked backers for "not only supporting my film, but for joining this MOVEMENT!" He encouraged backers to check out the social links to

Dear White People and to share the trailer with ten friends. The buzz generated by social media sparked attention from the news media.

In an interview with CNN, Simien said, "I'm trying to bring a new story to the screen," adding, "I think that we [Black storytellers] have a lot of stories to tell, and I think our point-of-view . . . can be universal." The online audience response to the trailer was overwhelmingly affirmative. Across the social media landscape—YouTube, Indiegogo, Facebook—the comments praised Simien's clever racial insights, cultural gaze, and brand of humor. The trailer's racial politics also stirred controversy. While some whites praised the concept trailer, several found it offensive and accused it of antiwhite racism. In an article Simien wrote for the *Huffington Post* promoting the trailer and the film he hoped to produce, he said, "My film really isn't about 'white racism' or racism at all . . . my film is about . . . identity. . . . "It's about the difference between how the mass culture responds to a person because of their race and who they understand themselves to truly be."

In the trailer, Simien deliberately channeled the cinematic auteurism and political opportunism of Spike Lee, a filmmaker he considered a role model for his ability to make commercial films about race while maintaining an art-house aesthetic. Hollywood's lack of interest in films that probe the complexities of Black life and identity created an opportunity for the provocative films Lee made. More than a quarter century later, Simien took advantage of similar opportunities.

The crowdfunding campaign was a success. A total of 1,318 people backed the campaign, which raised $41,405.

"Black audiences are thought of very myopically by Hollywood. It's been a very long time since there was an artful comedic approach to Black material," said Simien. "People were excited that it was something different than an inner city kid with a gun or a movie about slavery." Combining creative grit, a gift for racial satire, and the inventive use of social media, Simien, and the crowd that he tapped, were hacking Hollywood's contrived notions about what Black audiences were interested in seeing on the big screen.

The success of the YouTube trailer and the crowdfunding campaign gave Simien and his team more runway for production and the time and confidence to seek the financing necessary to produce a feature-length film. The campaign provided much-needed evidence to the filmmaker that an audience for his film—underrepresented and eager to see richer portraits of Black life on-screen—not only existed; it was determined to be heard.

@DEARWHITEPEOPLE: TWITTER AND
THE VIRTUAL WRITERS' ROOM

Simien's approach to refining his script included another novel use of social media. In order to develop and test ideas for the lead character and the film, Simien turned Twitter into a writers' room. In Hollywood, the writers' room is where a collection of talent develop the stories and characters that populate our entertainment enivronment. Using the Twitter handle @ DearWhitePeople, Simien began tweeting as Sam White, the lead character in the film. Sam's campus radio show, titled *Dear White People*, sparks some of the narrative tension in the film. Simien saw the two-way conversation on Twitter as an opportunity to develop Sam's character. "I had been working on the screenplay for some years at that point and wanted a way to sort of test out her very specific brand of humor on the world at large and see how it would react . . . both positive and negative."

Simien engaged in his own form of crowdsourcing through Twitter. More specifically, he used the social channel as an open network, both testing and soliciting ideas for dialogue and one-liners. In 2014, Simien told the online publication *Complex* that Twitter was "a way to see which of Sam's jokes would land and which would offend, and, if so, how they would offend. It was a research tool for me long before it became a promotional tool for the film." It was on Twitter that followers contributed some of Sam's most provocative lines in the film:

> Dear White People, listening to Flo Rida does not make you practically black.

There was also this one:

> Dear White People, the amount of black friends required to not seem racist has now been raised to two. Sorry, your weed man Tyrone does not count.

Twitter was also a great source of feedback. The platform helped Simien anticipate the avalanche of criticism from those who viewed the premise of *Dear White People* as antiwhite and, therefore, racist. For example, some of the cinematic conflict between Sam and the white students on campus first arose in the real world via Twitter. For instance, someone tweeted the following to @DearWhitePeople: "How would you feel if there was a 'Dear Black People?'" Another follower responded with this tweet: "Dear White

People, there's no need for a Dear Black People. Reality shows on VH1 and Bravo let us know exactly how you feel about us." It was a combination of tweets and micro-conversations like these that kindled a number of ideas for the script.

Simien's inventive use of Twitter as a writers' laboratory helped to establish a point of view for *Dear White People* that was unapologetically Black and distinctly millennial. Doing so not only sparked ideas for the film, it also helped to sustain the network of support that eventually made his vision of a feature film a reality.

TRAPPED BETWEEN MINSTRELSY AND MISERY

When Simien began pitching his movie to studios, he soon realized that his options were limited and his odds long. His successful concept trailer and crowdfunding campaign opened up studio doors, but they did not open up the minds of the executives he met. Studio executives liked the tone, look, and satirical motif of the trailer. But each meeting yielded the same result— no deal. In some cases, studios explained that they had no model for the film and therefore were unsure how to market *Dear White People*. Another key response was that the film was simply too different from the other "Black films" that had been greenlit by the industry. This was a not-so-subtle reference to the long shadow cast by Tyler Perry and his brand of entertainment.

What made selling *Dear White People* to studios especially hard from Simien's perspective was not only that it was an independent film not backed by any industry insiders but that it was an indie film featuring African Americans. Reflecting on his experience, Simien explains: "When you walk into a room to talk about financing, you find 96% of them won't even take a look at the project because it has black characters and a black cast and they 'don't think they'll be able to sell it internationally.' That is the definition of institutional racism." It also explains how exclusion limits the opportunities for new and different voices to gain a foothold in the industry.

After all of these years of moviemaking, Hollywood is still based on the "theory of precedence." That is, industry executives are more likely to make films that look like films they have seen before while also catering to the kinds of audiences that they are most familiar with. This has consequences for all filmmakers, especially those with stories that are not aligned with industry notions of what is bankable.

"Films with predominantly white casts can come in any form, tell any story, big or small," Simien told *Variety* in 2013. "For black films, you have the light fluffy rom-coms or Madea movies, and then you have the black torture awards movie." Young filmmakers like Simien have to reconcile the fact that their vision of quality entertainment typically falls outside of Hollywood's limited scope, which covers only the territory between minstrelsy and misery, between Perry's films and films like *Twelve Years a Slave* and *Precious*.

Millennial content creators face a stubborn reality: Hollywood remains resistant to diversity. After investigating more than nine hundred films made between 2007 and 2017, a study from the University of South California's Annenberg School of Communication found that "Hollywood is still so white." How white? Seventy percent of the characters in film are white compared to 14 percent Black, 6 percent Asian, and 3 percent Latino. Among the directors for the nine hundred films included in the study, 6 and 3 percent, respectively, were Black and Asian.

Rather than quit when the executives declined to finance *Dear White People*, Simien's team maintained an active social media presence. They kept the @DearWhitePeople Twitter handle live. In addition, Simien's team continued to release short videos on their YouTube channel that kept the conversation and the community active. A year and a half later the team had sustained a core audience of about 25,000 people who bought the merchandise, watched the online videos, and believed in the movie even when Hollywood did not.

Simien's persistence in feeding the creative beast eventually paid off. He secured financing for feature-film production from independent financier Julia Lebedev and her studio Code Red Films. *Dear White People* was also invited to the 2014 Sundance Film Festival. A total of 4,057 feature-length films from all over the world were submitted that year. The festival selected 118 feature films, representing thirty-seven countries and fifty-four first-time filmmakers. *Dear White People* was one of the most talked-about films at Sundance and also walked away with the Special Jury Award for Breakthrough Talent. The Sundance appearance produced significant tailwinds for Simien and his film, which appeared at several other prestigious film festivals before beginning theatrical distribution on October 17, 2014.

The boutique film did well in theatrical release, grossing $344,000 in eleven theaters opening weekend. The film grossed more than $4 million overall. Importantly, it played well with millennials. More than 80 percent of the audience was under the age of forty, and about 30 percent were young twenty-somethings.

None of this success would have been possible without social media and the new innovation economy. Simien's adoption of social media did more than establish a market for *Dear White People*; it also grew a community and an audible rebuke of Hollywood's biased view of Black moviegoers. The use of YouTube, Indiegogo, and Twitter amassed a substantial audience and a source of financial support that gave Simien, a novice filmmaking talent, a modicum of leverage despite the industry's inability to see the commercial potential of *Dear White People*. Simien describes this leverage: "It really, really helped that I never had to pitch it. I never had to start from zero. I had the trailer. I had a fan base. . . . So when I entered a room, people already knew who I was, because whether or not we were successful was not necessarily dependent on them. If they were coming on board, they were going to be hopping onto a moving train. And coming into a meeting with a moving train is very different than coming into a meeting expecting them to start the train."

THE INFLUENCE OF BLACK CROWD POWER

Justin Simien's and Issa Rae's stories are compelling, in part, because they would not have been possible just a few short years ago. Politically savvy web-based content like *The Misadventures of Awkward Black Girl* and *Dear White People* has transformed the pop culture landscape. These independently produced narratives radically expand our notion of what media entertainment can and should be while pointedly reproaching the lack of diversity among Hollywood content creators.

A recurring theme in my fieldwork is the transformative role of technology in the lives of young creatives. But this is not a story about what technology does to young creatives. Rather, it is a story about what they do with technology—innovation. Young creatives are transforming technology into vibrant innovation in the arts, media and entertainment, education, design, and civic life. Rae and Simien used social media as an innovation lab, a place where they could make their story ideas tangible and their aspirations possible. Whereas Rae used YouTube to prototype a television show from an unlikely point of view, Simien used the video channel to distribute a trailer for a movie he had not even made. In their engagement with social media, Rae and Simien identified and capitalized on a generational divide of which Hollywood executives were simply not aware—the growing displeasure among Black millennials with the limited scope of representations in entertainment media.

But these stories are not only about inventive content creators; they are also about the formation of powerful networked communities made possible through the internet.

There is a widely popular view that the internet democratizes our media environment by creating opportunities for a greater diversity of content creators. What also makes the web such a disruptive force in our culture is the growing number of people around the world who are building communities in the virtual territory they have staked out on the internet. As a result, the web's influence has come as much from the rise of connected communities as it has from the rise of content creators. As the population of people using the internet has become more diverse, the communities, conversations, and interests percolating across the web have also become more diverse.

Rae and Simien were the beneficiaries of an internet that allowed them to distribute, test, iterate, and raise capital for their ideas of entertainment despite Hollywood's disinterest. But they were also the beneficiaries of networked communities that demonstrated a demand for the kinds of stories the two of them were inspired to tell. Speaking to an audience of young aspiring media creators in 2014, Rae acknowledged the growing impact of African Americans online in her unlikely transition from the web to a premier cable channel: "If it weren't for social media, I don't know if we would even be a full form blip on the radar." She added, "Online content and new media have changed our communities and have changed demand and accessibility of content."

Rae and Simien leveraged Black crowd power to jumpstart their careers in media entertainment. More precisely, their rise in Hollywood points to the dynamic ways in which Black millennials mobilize online communities that are social, vocal, political, and skillful in the ways they use social media and crowd power to assert greater influence in the arenas of pop culture and politics. This application of social media has become a subtle but crucial feature of the new innovation economy.

CHAPTER 7

STEM GIRLS

Expanding the Talent Pipeline in the Tech Economy

Every spring and summer Silicon Valley goes through what has become an annual ritual since 2014. The tech companies there release their workforce diversity statistics for the world to see. Google was the first tech company to release their data in 2014. The tech companies had consistently refused to share their workforce stats, claiming, among other things, that they were a trade secret. This was always a curious excuse and one that over time became increasingly difficult to defend. Tech activists continued to pressure the companies for several reasons.

First, Silicon Valley has become enormously wealthy and now asserts an outsized influence on the larger economy. In 2018 Apple became the first company in history to reach a trillion-dollar valuation and ranks with Google, Amazon, and Facebook among the most valued companies in the global economy. Second, there is growing recognition that some of the best paying and most secure jobs in tomorrow's economy will be in science, technology, engineering, and math (STEM). According to the Bureau of Labor Statistics, STEM employment will far outpace non-STEM employment in the expected rate of job growth. Whereas the projected growth rate for all occupations between 2016 and 2026 is 7 percent, the growth projections for STEM occupations like mathematical science and computer occupations are notably higher at 28 percent and 12 percent, respectively. Already, STEM workers, on average, earn 26 percent more than their non-STEM counterparts. Amazon, Apple, and Google each made news in 2018 by announcing the expansion of their operations in cities as varied as Arlington, Virginia; Austin; and New York. Facing political pressure, Amazon pulled out of New York.

Finally, Silicon Valley has become a central player in just about every aspect of our lives. Several of the products and services that tech companies build influence how people around the world connect, engage in civic life, learn, consume media entertainment, and manage their private affairs, including their health and well-being. As the reach and power of big tech companies has increased, the gender and racial disparities in hiring and promotion have become conspicuous and troubling.

In 2014, 70 percent of all Google employees were male, and 91 percent were either white or Asian. Facebook's workforce that year was 69 percent male and 91 percent white or Asian. Yahoo reported that men made up 62 percent of its overall global workforce. In the US, 89 percent of Yahoo's workforce was either white or Asian. Microsoft's workforce was 72 percent male and 90 percent white or Asian. Other companies including Oracle, LinkedIn, Pinterest, Amazon, and Twitter released similar workforce numbers. The data was clear. The tech industry suffers from what can only be described as a diversity crisis.

In 2016 a graduate student and I examined the tech workforce data closely and concluded that the diversity crisis persists. In our analysis, we identified companies whose lack of gender diversity reflects a condition we call *gender hyper-exclusion*, a reference in this instance to companies whose employment of men in leadership or technical positions is at or above 70 percent. Gender hyper-exclusion accurately describes an extreme imbalance between the proportion of men and women working in tech.

Many of the top tech companies in the US reflect gender hyper-exclusion in their leadership and technical staff, including Airbnb, Amazon, Apple, Facebook, Google, Intel, Microsoft, Pinterest, and Yahoo. Gender hyper-exclusion is even more pronounced in the highly coveted technical positions like engineering, programming, and computer network architects, where women are virtually shut out.

Several factors contribute to the diversity crisis in tech. The organizational behaviors of tech companies are certainly a contributing factor. Industry practices like unconscious bias and recruitment bias are notable culprits. Unconscious bias, in this context, references the ways organizations make decisions without reflecting on the machinations of those decisions. Decisions, for example, about hiring, promotion, and compensation are often made according to organizational norms, cultural beliefs, routine procedures, and gut instincts that produce outcomes that, for example, favor male employees over their female counterparts. Many tech companies now conduct bias-busting workshops and other training sessions to encourage the recognition and analysis of unconscious behavior.

Furthermore, some tech companies are examining how they recruit talent. Each year the tech companies scour the nation's college campuses to find young talent in engineering, computer science, and design. A few years ago Google and Facebook began to notice something in particular about *where* they recruit. More precisely, they recognized that their annual recruitment focuses on elite college campuses, that is, campuses that are likely to have only a small percentage of women, Black, and Latino graduates in areas such as computer science and engineering. In 2016 Google and Facebook launched more assertive initiatives to connect with historically Black colleges and universities to widen the pool of talent that they evaluate and recruit from. The idea is simple but powerful: if you do not look in the places where more diverse talent is located, the likelihood of building a more diverse workforce will be severely limited.

And then there is the tech talent pipeline. Among the underlying factors driving the diversity crisis, none may be more urgent than the STEM readiness crisis caused by an educational system that does not prepare girls and young women or Black and Latino students for meaningful participation in the STEM economy.

THE STEM READINESS CRISIS

One of the main pathways to a successful career in STEM is the attainment of a college degree in a STEM field. In 2013 and 2014, 1.8 million bachelor's degrees were awarded to US citizens. Of those degrees 17 percent were in STEM fields. Predictably, the distribution of STEM degrees varied significantly by race and gender. For example, the percentage of Asian students receiving a STEM degree, 31 percent, was nearly double the percentage awarded to students overall. White students earned STEM degrees at the same rate that such degrees were awarded to students overall, 17 percent. By contrast, the percentage of STEM degrees awarded to Latino and Black students, 14 and 11 percent, respectively, was lower than the percentage awarded to students overall.

There is gender variation in STEM degree attainment too. Throughout much of the twentieth century, men earned more bachelor's degrees than women. But women started closing the gap in the middle to late 1970s. By 1980 half of the bachelor's degrees conferred in the US went to women. Since then, women have earned a growing share of the bachelor's degrees conferred in the US. By 2014 women were earning more than half, 57 percent of all bachelor's degrees conferred. However, when it comes to STEM degree attainment, women fall far behind men.

In 2013 and 2014 women earned 35 percent of the STEM degrees in the US. Women were much less likely than men to earn a STEM degree across all racial and ethnic groups. Asian women earned 40 percent of the STEM degrees attained by Asians. White women were the only female group to earn a percentage of STEM degrees, 33 percent, that fell below the percentage awarded to women overall, 35 percent. Black and Latino women earned, respectively, 44 and 36 percent of the degrees awarded to their racial and ethnic group.

The disparities in STEM degree attainment are the results of an education system that, historically, has failed to educate students equally. In truth, the STEM readiness crisis begins long before young people enter college. Most education scholars believe that learning disparities begin to form during the early years of schooling. Some of the most influential work in this area is by the Nobel Laureate economist James Heckman. According to Heckman, who is an economist at the University of Chicago, early childhood cognitive ability is a strong predictor of future academic, social, and economic success. Heckman has spent much of his career studying the long-term effects of the cognitive disparities that form in early childhood. He argues that skills gained early in life—cognitive and noncognitive—form the foundation for skills later in life. Those who acquire certain skills as children continue to build on them over the course of their lives. Education scholars call this the "Matthew Effect," from the biblical adage "the rich get richer and the poor get poorer." We can infer from Heckman's research that the STEM racial and gender learning gaps that become so apparent in high school and college likely begin in early childhood.

Consider, for example, fourth- and eighth-grade math performance on the National Assessment of Educational Progress, generally regarded as the nation's report card. According to the National Center for Education Statistics, the average fourth-grade math score in 2015 for white students was 248. By comparison, the average fourth-grade math scores for Latino and Black students were 230 and 224, respectively. Asian/Pacific Islander students had the highest average test score, 257. These gaps increase as students progress through high school. The eighth-grade math score for white students was 292 in 2015, compared to 260 for African American students and 270 for Latino students. Among Asian/Pacific Islander students, the average eighth-grade score in math was 306. Notably, the gap between whites and their Latino and African American counterparts widens between fourth and eighth grade. The gap between Asian students and their Latino and Black counterparts also increases from fourth to eighth grade.

The gap in math test scores between male and female students is not nearly as substantial as the racial and ethnic gaps. In 2015 the average math score for fourth-grade male students was 241 compared to 239 for female students. The average math score among eighth-grade boys in 2015 was 282, which was equal to girls that year. Despite test scores that are virtually equal, boys often express higher confidence than girls in subjects like math.

A study by education researchers at Florida State University found some interesting differences between boys and girls in STEM subjects like mathematics. "The argument continues to be made that gender differences in the 'hard' sciences is all about ability," said Lara Perez-Felkner, assistant professor of higher education and sociology in Florida State University's College of Education and lead author of the study. "But when we hold mathematics ability test scores constant, effectively taking it out of the equation, we see boys still rate their ability higher, and girls rate their ability lower."

Perez-Felkner's research team found that the perception gap actually widens at the upper levels of mathematic abilities, that is, among boys and girls with the highest skill level in math. More specifically, she found that high-ability boys are more confident than high-ability girls when working in challenging mathematics contexts. We can extrapolate from this research that even when girls have the potential to thrive in STEM learning contexts, they generally lack the confidence that carries boys through the inevitable rough patches that crop up with challenging material. The researchers maintain that this lack of confidence affects the classes girls choose through high school and the course of study they pursue in college. Even when women have the ability to succeed in STEM, they may not stay in the pipeline because they don't feel they can succeed there.

The next three chapters focus on the growing number of young creatives setting their sights on the STEM readiness crisis and the lack of diversity in tech. Whereas public schools have largely been unable to deliver high-quality education, young creatives, including designers, educators, and social entrepreneurs, are prototyping new learning futures that hold the promise of building a more diverse pipeline of tech talent. In this chapter, I consider the work of Debbie Sterling, the founder of GoldieBlox, a multimedia company that seeks to empower girls to succeed in engineering. In chapter 8, I profile Black Girls Code, a nonprofit organization that introduces Black and Latina girls to the principles of computer programing. Finally, chapter 9 considers the Qeyno Group, an Oakland-based entity that organizes hackathons for Black teens.

LONELY GIRLS: WHY FEMALES STRUGGLE IN STEM EDUCATION

The US workforce has changed in some extraordinary ways over the last half century. One of the most notable changes is the higher percentage of women participating in the paid labor force. In 1950 women accounted for about a third of the labor force. By 2015 women represented about 47 percent of the overall labor force. But even though women play a much more significant role in the jobs economy today, they still suffer from a number of disadvantages compared to their male counterparts. The Bureau of Labor Statistics reports that in 2016 women who were full-time wage and salary workers had weekly earnings that were 82 percent of the earnings of their male counterparts. The earning disparities between women and men vary by occupation. For example, women in engineering occupations earn 79 percent of what their male counterparts earn. A female software developer earns 83 percent of what her male counterpart earns.

Moreover, even as women are earning more bachelor's degrees than their male counterparts, 57 percent to 43 percent, they are still limited in terms of their occupational choices and mobility. We continue to associate certain professions with gender. Try this: visualize a female worker in your mind. You likely see images of a secretary, nurse, or teacher. Now visualize a male worker. Despite a rapidly changing economy, it is likely that images of a factory worker, lawyer, or construction worker come to mind.

Because these gendered differences are so firmly etched in our minds, it is easy to think that men and women are just innately better at certain occupations than others. We see these gendered occupations as natural—even biological—rather than what they really are, cultural and sociological constructs. Men are not naturally better surgeons; they are just more likely than women to be socialized to see themselves as surgeons. This kind of gender socialization is especially apparent in STEM occupations.

Debbie Sterling, the founder of the company GoldieBlox, came face-to-face with this dilemma when she enrolled in Stanford University in the early 2000s and elected to major in mechanical engineering. As a high school senior in a small Rhode Island town, Sterling asked her math teacher to write her a letter of recommendation for her college application. When the teacher asked what she wanted to major in, Sterling was not really sure.

The math teacher replied, "How about engineering? I think you would really excel in it."

When Sterling visualized an engineer, she pictured a man. A train driver to be exact, she told an audience in a 2013 talk. She had no idea what an engineer was. She asked herself, "Why would a creative, artistic girl like me want to be an engineer?" Sterling did not hesitate to pursue an engineering degree because she was not smart enough. Rather, she never considered the profession because she had never seen an engineer who looked like her. As research from social scientists suggests, it is hard to become something for which you have no role models. It was not surprising that Sterling had never seen a female engineer, considering that when she began college, women made up only about 11 percent of the engineers in the US.

Because she trusted her math teacher, Sterling decided to major in engineering at Stanford. One of her first classes was Mechanical Engineering 101. She did not have a lot of confidence when she began the class. Making matters worse was that there were so few women in the course. But, much to Sterling's surprise, she loved the course. "It was so cool, and so much fun," Sterling recalled. "In that class I learned that engineering is the skill set to build anything you dream up in your head. Whether it's a website, a mobile app, or a bridge to a highway."

In her first engineering course Sterling experienced things she had never experienced before in a class: excitement and possibility. These feelings, however, were tempered by another experience: the feeling that she was alone. The low ratio of women to men in her engineering classes took a subtle but notable toll on her. No matter how much she tried to get over it, she constantly felt as though she did not belong. Engineering, she thought to herself, was for men, not women.

Sterling's experience at Stanford was not an isolated case. In fact, it continues to be a common experience that women in STEM majors experience the feeling of being alone, usually because they are.

Nilanjana Dasgupta, a psychology researcher at the University of Massachusetts Amherst, has collected empirical evidence suggesting that a key predictor of women earning a STEM degree is how well they grapple with feeling alone or isolated in their major. "Belonging," Dasgupta told the National Science Foundation, "determines whether you stick it out in a field that interests you." Most women in STEM majors experience a lack of camaraderie and comfort. As a result, some begin to lose interest and confidence, according to Dasgupta's research. "Belonging is just a way of saying 'Do I fit in here? Do I feel comfortable here? Or should I start looking for another subject where there are more people like me?'"

The University of Massachusetts professor maintains that most people think of performance as determining whether someone chooses an academic pathway or career. She maintains that performance—or the ability to do STEM academic work—is not what she looks for when assessing whether or not women will persist and earn a STEM degree. The women who leave a STEM major for another major tend to perform just as well as the women who stay in the major, according to her research. Dasgupta says that "feeling like they fit in, or not, is the critical ingredient that determines retention."

Debbie Sterling, recalling her Stanford days in front of an audience, shares just how haunting the environment was. Halfway through her major, she took an engineering drawing class. "Great," she says, "I would get to draw." But she soon found out that students were expected to draw in perspective, that is, in 3-D. She worked hard in the class, but for some reason the harder she worked, the more she seemed to struggle. For the final class assignment, students were asked to display their drawings. According to Sterling it was obvious that many of her male classmates had scribbled their drawings minutes before class and "slapped them on the wall." Sterling added, "Meanwhile, I had spent hours, the entire weekend, working on my drawing."

As the instructors approached her drawing, they turned to the class and one of them said, "Raise your hand if you think Debbie should pass this class." A few students cautiously raised their hands. Sterling felt humiliated. A male student came to her defense, excoriating the instructors and telling them that it was their job to teach her, not to shame her. Horrified by the experience, Sterling burst out of the classroom in tears, telling herself, "This isn't for me. I'm not naturally good at this stuff." She considered giving up engineering, but the friend who had come to her defense encouraged her not to quit the major. Eventually, in spite of this experience and others like it, Sterling composed herself and went on to earn her degree in mechanical engineering from Stanford.

CHILDHOOD PLAY AND THE STEM TALENT PIPELINE

Like most young graduates, Sterling started her career with a series of jobs that paid her modestly. She began as a lowly paid intern at a design consultancy in Seattle but worked her way into a full-time position. Sterling would go on to other jobs and even worked for a nonprofit for a year in India. As was the case with many of the young creatives that I met, Sterling longed for more.

While working in the Bay Area, Sterling, along with some people she knew, organized an informal club called "Idea Brunch." "Once a month, we'd get together, cook a big breakfast, and each person would get up in front of the group and share their latest idea," Sterling recalled in an interview with *Forbes*.

Throughout my research I discovered several instances of young creatives organizing informal events similar to this one. A group I learned about in Detroit gets together over soup and salad to pitch their ideas for new enterprises. Though the groups and the locations vary, the idea is generally the same: young creatives getting together periodically in an intimate social setting to offer support, a listening ear, or friendly companionship. These informal meetings are an opportunity for participants to share new projects or ideas. In other instances participants might have a chance to get feedback on a new start-up or nonprofit. Importantly, these gatherings are yet another example of how young creatives are designing inventive ways to cultivate personal relationships that grow their social capital and feed their entrepreneurial ambitions.

During one Idea Brunch, a former classmate of Sterling's recalled their days as engineering students at Stanford. "My friend, Christy, . . . got up and started complaining about the lack of women in our classes," Sterling said. Christy mentioned how as a kid she had played with her brother's construction toys. She believed the toys had sparked her interest in making things and had given her the impetus to eventually pursue a degree in engineering. According to Sterling, her friend's idea was "engineering toys for girls." The idea struck a chord with Sterling immediately. Several weeks later she was still thinking about a construction toy for girls.

Sterling became curious about the link between how we are socialized as children and the pathways we take later in life. She began conducting research and soon discovered that the kids who score better on spatial skills test are more likely to have grown up playing with construction toys like Legos and erector sets. Maybe, she thought, that explains why so many young women struggle in certain engineering courses—they didn't grow up playing with toys that developed spatial skills. She had always believed that her difficulties in the engineering drawing class could be attributed to the fact that she was not as naturally gifted as her male counterparts. In other words, she had attributed her performance in engineering drawing to biological factors. But the research she discovered suggested that her struggles were more likely attributable to the fact that girls, compared to boys, are less likely to be exposed to forms of early childhood play that develop their spatial skills. If this was true, she reasoned, then a correction of some kind in the cultural environment could strengthen young girls' performance in academic settings that required these skills.

Sterling discovered what researchers have long understood about the power of play. Through play, children reenact certain established narratives about gender, race, class, and identity. The roles they try on during play become the frames through which they imagine themselves as adults, their future selves. Through play, children develop the scripts, norms, and ideologies that inform their notions of who they can be in the world. But play is also socially constructed, shaped by beliefs, ideas, and customs that we seldom explicitly recognize but reproduce via some of the main institutions in society, such as family, school, and pop culture. Play is an incredibly powerful experience that has cultural, behavioral, and ideological significance. Still, we largely take play for granted and consider it inconsequential.

After learning about the power of play and gender socialization, Sterling decided to build a construction toy for girls that introduces them to skills like tinkering, building, and problem-solving. She was an engineer by training, but she was also developing the skills exhibited by young creatives. She approached the project by thinking like a mechanical engineer, certainly, but also as a designer and an educator. Sterling did not necessarily know it at the time, but she was contemplating a radical disruption of the status quo that had been more than five decades in the making. In 1957 another young female entrepreneur named Ruth Handler broke social barriers by creating a toy for girls and naming it after her daughter, Barbara. The toy, Barbie, transformed the toy industry and how generations of girls played and fashioned their social identities.

REINVENTING GIRL'S PLAY

When Sterling elected to create a construction toy for girls, she called into question a multibillion-dollar toy industry premised on long-standing ideas about gender identity, society, and children's play. She conducted research on children's toys, sharpening her insights about a variety of things, including shapes, colors, and styles of play. Sterling bought paper, plastic, tape, glue, and fabric from the hardware store and began experimenting with prototypes. Her small apartment was filled with various samples of her early ideas.

Part of her research also included ethnographic inquiry. For instance, she met with little girls to begin testing her ideas and learning more about what types of toys and play experiences resonated with them. This fieldwork led to a creative breakthrough: Sterling noticed that when the girls played with construction toys, they would get bored and stop. One day when an eight-year-old girl lost interest in one of the toys available, Sterling asked what her favorite toy was. The young girl grabbed a book and asked if they could read it together. This exchange with the little girl sparked an aha moment later that day, when Sterling thought, "What if I put those two things together, spatial plus verbal, a construction toy plus stories?"

She immediately started sketching characters and story scenarios to go along with the toy kit. The stories would feature a spunky girl innovator named GoldieBlox and her many adventures. With each adventure, GoldieBlox solved problems by building solutions. The girls could read and play along and build things similar to those GoldieBlox was building in the story. This new concept was an opportunity for Sterling to fulfill her initial desire to introduce young girls to tinkering and making—STEM role modeling—while also engaging them in a form of play—storytelling—in which many of the girls expressed an interest. Sterling used her background in engineering and design to prototype the new concept. Like a good designer, she tested the prototype toy kit with young girls around the Bay Area.

The concept was a hit. The girls loved the stories, which gave them additional insights into GoldieBlox as an inventor while also catalyzing their play, imagination, and maker skills.

KICKSTARTING A CONSTRUCTION TOY FOR GIRLS

As Sterling considered how to gain access to the toy industry in order to get GoldieBlox manufactured and on the market, a friend persuaded her to apply for admission to an elite tech accelerator in Silicon Valley. But when she appeared before the all-male panel of tech bros, they did not understand why the construction kit needed a book. It was not simply that they did not get her toy, they did not get why a toy like this was worth their investment. Sterling left feeling much as she had while studying engineering at Stanford: as if she did not fit in.

Next, she presented her concept at the New York Toy Fair, an international event and the biggest toy show in the US. She was excited to attend, expecting to meet creative types, kids, and people who were investing in the future of toys. "What I encountered instead," she said, "was a bunch of old men in suits." Again, as was true during her days at Stanford and at the Silicon Valley accelerator, Sterling did not fit in.

As she made the rounds at the Toy Fair, several men shared a well-known industry secret: "Construction toys for girls don't sell," they told her. They explained that princess and Barbie dolls were what sold. They thought her idea was novel, even noble, but it was not commercial.

Despite her struggles to make her vision real, Sterling persisted. She partnered with a manufacturer to produce a sample toy. The first complete set she designed included a storybook, character figurines, and construction pieces—a pegboard, a spinning wheel, axels, blocks, ribbon, and washers in bright colors and with curved edges. The stories were intentionally fun, accessible, and humorous. The goal was to make building things engaging rather than rote and complicated.

In order to go to production, the factory minimum order was five thousand toys. Sterling needed $150,000 to move forward. Lacking that amount of money, she did what a lot of young creatives we have met do: she launched a Kickstarter. Her goal? Raise $150,000 in thirty days.

In her Kickstarter video Sterling asks potential contributors to "help me build GoldieBlox so our girls can build the future." The response was overwhelmingly positive. Sterling reached her fundraising goal in four days. Ultimately her 5,519 backers contributed a total of $285,851. The Kickstarter campaign allowed her to go into production and make her vision of a construction toy for girls real.

BUILDING THE KID TECH ECONOMY

Sterling's construction toy for girls was not only inventive; it was also timely. GoldieBlox came to fruition just as the movement to diversify the tech talent pipeline in the US was ramping up. As concerns about the lack of gender diversity in tech grows more vigorous, the momentum to boost the presence of women in engineering fields is growing. Whereas the tech bros representing the Silicon Valley accelerator and the old guys in suits at the New York Toy Fair failed to see the promise in GoldieBlox, the internet crowd got it. The Kickstarter campaign and a series of viral YouTube videos tapped into a community of advocates and influencers who became evangelists for

GoldieBlox. In less than two years Sterling, a toy industry outsider, went from an inspirational idea to drawings in a sketchbook to homemade proto-types to a manufacturing order to online sales to a Super Bowl commercial and, finally, to the shelves of Toys "R" Us.

Right around the time GoldieBlox was beginning to establish itself as a unique product the toy industry was grappling with its own tech-inspired disruption. More specifically, a burgeoning economy around apps, mobile games, and interactive storytelling was emerging and reshaping the future of children's play and learning. Today, more of the toys that young girls and boys spend their time with are digital and interactive. The demise of Toys "R" Us is a clear example of an industry in transition. The toy retailer filed for bankruptcy in 2017. In 2018, the company announced that it was closing its two thousand US stores as internet retail and a new generation of toys combined to upend what the *New York Times* called "the iconic retail chain that has sold toys and games to millions of children for generations."

The shift of children's play to smartphones and tablets is a source of great alarm for critics who believe that digital gadgets undermine the de-velopment of young children's literacy. In 2018 a series of reports in the *New York Times* note how Silicon Valley parents, among others, are severely restricting children's access to screens. The truth is that researchers do not fully understand the implications of the digitization of children's play and literacy environments. While educators, policy makers, pediatricians, and children's organizations usually aim their concerns at what is called "screen time," that is, the amount of time that kids spend looking at screens, the more substantive issue is the quality of the content that kids engage with during their time with screens.

In 2015 I and some other researchers and children's media profession-als were invited by the American Academy of Pediatrics to a meeting with the intention of updating their recommendations to pediatricians on what constitutes a proper media diet for children. Pediatricians are besieged with questions from parents seeking guidance regarding what constitutes a healthy media environment. For years the AAP has recommended that par-ents limit their kids' screen time to two hours a day. But as any parent knows, that is not feasible. Much of what we do in our everyday lives involves a screen. This is increasingly true for the youngest members of our society too. The books that kids read, the games that they play, the media that they watch, and even the schoolwork that they do is increasingly mediated via a screen, usually a tablet, smartphone, or computer.

Some of us attending the meeting suggested that the AAP focus less on the amount of time children spend with screens and instead encourage parents to be more attentive to the quality of the content their kids consume. Researchers also encourage parents to practice co-viewing, which is when parents watch or participate in media with their children. There is an emerging body of research that suggests that co-viewing can help emphasize the learning opportunities embedded in children's media.

Just a couple of years after breaking through the pink aisle at Toys "R" Us, Debbie Sterling and the GoldieBlox team began a strategic shift to transform into a multimedia company. "We've launched multiple YouTube series, released our first chapter-book series with Random House Children's Books and created STEM engineering badges for the Girl Scouts," Sterling told *Forbes* in 2017. The shift is based, in part, on the multiplatform world that is driving the media and entertainment economy. The company believes that the time is now for apps, games, and interactive stories that develop STEM literacy among girls. Beau Lewis, VP of content at GoldieBlox, told Kidscreen that the goal of the GoldieBlox YouTube series is "to actually result in learning and offline exploration for kids. . . . We created these inventions around relatable kid problems, like trying to get candy off of the top of the refrigerator. We put a lot of focus on making those inventions whimsical and creative to show the fun side of engineering."

The GoldieBlox YouTube channel reflects the range of opportunities made possible in the new media ecosystem powered by social media. YouTube has been especially assertive in growing their division geared toward young audiences. The company has stepped up its production of content for children, enlisting a whole new generation of kids to see YouTube as their go-to source for media and entertainment. But some critics are raising questions about YouTube's maneuver to connect with children.

YouTube has come under fire for allegedly collecting data on children under the age of thirteen, marketing consumer goods to them, and allowing inappropriate content to slip through their algorithmic filters on the YouTube Kids app. In 2018 a coalition of twenty children's advocacy groups announced that they were going to file a complaint against YouTube to the Federal Trade Commission. The coalition charged that YouTube was illegally collecting data on young children and profiling them, a violation of the Children's Online Privacy Protection Act. Claims like these point to larger questions about what has been called "kid tech" and the ethics and impact of tech companies' growing role in the lives of children.

As one executive of a children's advocacy group told the *New York Times*, "Silicon Valley now has a responsibility to figure out how to move from building adult tech into building kid tech," he said. "I don't think any company, whether it's Facebook or YouTube or Snapchat, can hide from the fact that there are about 10 times more kids online today than there were six or seven years ago." The rise of kid tech signals the steady encroachment of artificial intelligence into children's daily lives through toys, digital assistants, tablets, and smartphones. Kid tech is here, but what does its future hold? In other words, how do we design kid tech in ways that reflect serious engagement with the ethical and behavioral issues that are intricately involved, for example, with children interacting with conversational AI platforms? What values will drive the design of the immersive media experiences that are poised to shape the future of childhood?

Young creatives like Debbie Sterling and the savvy education start-ups they launch will and must be significant players in the kid tech economy. Approaching kid tech from outside the tech industry, they are much more likely than profit-driven corporations to see it as a world in which the smart design of apps, games, interactive stories, and mixed reality can be the building blocks of a new approach to children's media, entertainment, and educational development rather than a data-rich terrain to be exclusively mined, corporatized, and monetized.

CHAPTER 8

CODE FOR CHANGE

Who Will Build the Smart Future?

Companies like Google, Facebook, and Amazon have gained enormous wealth and cultural influence due, in large part, to their ability to design computer-based platforms that collect, mine, and exploit the data generated by their users. These companies are governed by algorithms or computer-based mathematical instructions designed to solve problems. Tech companies use algorithms, for example, to create behavioral profiles based on billions of people's internet use and then, using predictive analytics, offer them certain content, advertisements, and services. Algorithms grant tech companies unprecedented knowledge about human behavior in the connected age. In recent years a robust public debate has emerged about the social and ethical aspects of algorithms and the increased presence of artificial intelligence in our daily lives.

Algorithms drive our society and economy. These seemingly innocuous pieces of computer code are extraordinarily powerful. They shape the information we see when we conduct a search, the ads that vie for our attention whenever we log into a device, and the content we are exposed to when we engage social media. Algorithms are also influencing the way institutions—including the judicial system, big retail, real estate brokers, and even politicians—engage in micro-targeting, or using predictive analytics to direct specific content to specific groups with unprecedented precision.

There is growing evidence that algorithms reproduce social and economic inequality too. For example, judges across the US are using an algorithm called COMPAS to issue prison sentences that raise new questions about race, class, and the criminal justice system. Algorithms help to explain why women are more likely to see ads for lower paying jobs and African Americans for stores and houses in lower-valued neighborhoods. Politicians and political consultants are deploying algorithms to micro-target potential voters with ads, campaign videos, and even "fake news." Algorithms do not just process one person's data and use it to predict that person's behavior. Algorithms are trained to make decisions about an individual based on how people similar to them in terms of race, gender, class, and geography, for example, have behaved.

As our knowledge grows about how companies like Facebook and Google deploy algorithms to rule our digital society and our economy, concerns are intensifying regarding the human decision-making processes that inform the artificial intelligence systems that govern our data-driven world. Algorithms are not neutral. They reflect the biases, explicit and implicit, of the people who design and train them. The fact that the algorithms that govern the digital world are largely designed by a homogeneous group of tech managers and computer scientists—white men—is emerging as the new ground zero in the battle to build an internet that is more equitable and reflects the global population that uses it.

For much of the internet's early life, society demonstrated little if any skepticism about how the technology works and shapes our lives. Tech companies, in fact, operated with little if any societal resistance or scrutiny. Google, Facebook, Amazon, Apple, and Microsoft were widely celebrated for their innovative products and services. But as the internet has evolved into, among other things, a global force in cultural and political life, problems have surfaced in the areas of personal privacy, corporate ethics, network security, data rights, and organized disinformation campaigns. The lack of diverse voices and expertise around the table when algorithms are being designed, developed, and deployed is simply no longer feasible. The stakes are higher than ever before when it comes to building a more diverse pipeline of talent in the fields of expertise that will shape the smart future, like computer science, design, and psychology.

WHO GETS TO LEARN COMPUTER SCIENCE?

In the previous chapter I discussed the diversity crisis in tech, including the extreme gender imbalance in employment in management and technical positions—gender hyper-exclusion—that my graduate research assistant and I found. Our research also identified companies whose lack of diversity reflects *racial hyper-exclusion*, the employment of whites and Asians combined in leadership or technical positions at or above 90 percent. Racial hyper-exclusion is clearly evident in terms of leadership positions. Some of the companies that met or exceeded the 90 percent threshold in their leadership hires included Airbnb, Amazon, Facebook, Google, Intel, LinkedIn, Microsoft, Pinterest, and Yahoo. Our analysis of workforce data suggests that in many companies racial hyper-exclusion affects the hiring of tech workers as well. Key players, including Airbnb, Amazon, Cisco, Facebook, Google, LinkedIn, Microsoft, Pinterest, and Yahoo, reported that 90 percent or more of their technical employees are white or Asian.

As we see with gender disparities in STEM employment, racial disparities can be linked to several factors, including the STEM readiness crisis in our schools. The lack of Black and Latinos in technical positions in tech companies can be attributed, in part, to the fact that only a small percentage of these students have access to high-quality STEM learning opportunities. Consider, for example, the sharp racial/ethnic and gender variations in computer science. A good proxy for access to high-quality instruction in computer science in high school is the Advanced Placement Exam in Computer Science. Students who take the exam have likely taken courses in computer science and, thus, developed knowledge and experience in the subject.

In 2017 those who took the exam in computer science were overwhelmingly male and white. The percentage of females that took the AP CSA exam in 2017 was 24 percent. Generally, the majority of AP exams are taken by females, a fact that makes the low percentage of females taking the test in computer science even more conspicuous. The racial and ethnic breakdown suggests that Latino and Black students have extremely limited access to computer science courses. The percentage of Latino test takers for 2017 was 12 percent. Four percent of the exam takers were African Americans. Latinas and Black females were the lowest percentage of any group in the US to take the AP Computer Science test.

A decisive majority of US high school students do not take computer science courses. According to the 2015 National Assessment of Educational Progress's (NAEP) grade-twelve student survey, just 22 percent of students reported taking a course in computer programming while in high school. A 2016 study by Google and Gallup asked high school principals why their schools do not offer computer science courses. Several principals explained that they did not have the curricular resources or instructional expertise. Half of the principals said they did not offer computer science courses due to a lack of time in their class schedule for subjects that were not part of the testing requirements. It is a sad commentary on the state of the education system that standardized testing, in some cases for skills that are not relevant in a knowledge economy, takes precedence over more relevant skills like computer science.

In my own research I have observed how the lack of access to rich STEM opportunities also impacts the tech talent pipeline in the US. The research, for example, found that the schools that Latino, African American, and lower income students attend consistently lack teachers who are qualified to teach grade-level skills in STEM; financial resources to upgrade their instructional expertise; and the curricular resources and infrastructure to support learning that prepares students for college-level STEM course work. Our research team found evidence that the academic achievement gaps in STEM are less about access to technology and more about access to high quality STEM learning opportunities.

The Bureau of Labor Statistics projects that employment in computer and information technology occupations will grow 13 percent from 2016 to 2026. That is nearly twice the projected growth of all occupations combined, 7 percent. Some occupations like software developers (24 percent) and computer and information research scientists (19 percent) are expected to grow at an even faster rate. In total, computing occupations are projected to add 557,000 new jobs over a decade. A surging number of these jobs will be heavily influenced by the spread of artificial intelligence in our lives. Virtually all of these jobs require at least a bachelor's degree, in addition to skills related to design, technical expertise, and analytical thinking. But given the low number of women, Blacks, and Latinos in the computer science pipeline, this growth sector in the economy will largely be off-limits to them.

Some argue that the language of the twenty-first century is not English, Mandarin, Arabic, or Spanish but computer programming. In the US some states have even gone so far as to credit computer science as a foreign language in their academic requirements for graduation. And while there are certainly limits to that kind of logic, it illustrates the significance attached to the mastery of computer science skills. In recent years, organizations like Code for America, high-profile tech leaders like Bill Gates and Mark Zuckerberg, and the Obama White House have become advocates for democratizing access to computer science education. Equity in STEM has become an opportunity space for innovation in education. Still, schools lag far behind the call to make computer science education for all a reality. A new generation of educators, designers, and social entrepreneurs are mobilizing their resources to fill this void and build a more diverse talent pipeline of computer scientists. One of these social entrepreneurs is Kimberly Bryant, the founder of a nonprofit called Black Girls Code.

Bryant's own background in computer science is interesting. The Memphis native attended Vanderbilt University, where she took her first computer programming course her first year in college. According to Bryant, "Few of my classmates looked like me." She enjoyed her classes and her classmates, but she also felt culturally isolated. "While we shared similar aspirations and many good times, there's much to be said for making any challenging journey with people of the same cultural background," Bryant recalled years later.

The idea that a computer science major would be introduced to a computer science course in college is highly unlikely today. Many students interested in computer science have likely been exposed to the skills associated with the field prior to starting college. The most advanced students enter college with proficient computer programming skills and are even placing out of introductory courses in computer science. The data, however, suggests that Blacks, Latinos, and women are much less likely to have adequate preparation to succeed in college-level computer science course work.

Before starting Black Girls Code, Bryant was one of the few Black women in the biotech industry, making her part of a select club. But her select status was not enviable. She was always one of the only women or African Americans on the teams that she worked on. When she traveled to technology conferences, Bryant rarely saw Latinas or other Black women. Over the years she felt increasingly isolated despite her personal success. When her young daughter expressed an interest in technology, Bryant discovered that there were few opportunities for her to learn coding skills. The opportunity for her daughter to learn coding from people who looked like her was even less likely. Experiences like these planted the seeds for her social enterprise in computer science education.

BLACK GIRLS AND STEM

The premise of Black Girls Code is straightforward. At its core the idea is to begin building a generation of diverse and talented young girls who can imagine people who look like them designing the smart technologies of tomorrow. Black Girls Code provides out-of-school learning opportunities for girls age seven to seventeen to develop basic skills in computing areas like HTML/CSS, robotics, and game and mobile app development.

The name notwithstanding, the organization is open to all girls but emphasizes participation from African Americans, Latinas, and Native Americans. Women in general are underrepresented in tech, but African American, Latina, and Native American women suffer even greater degrees of hyperexclusion. In 2016 women received 57 percent of the bachelor's degrees awarded in the US but only 19 percent of the degrees in Computer and Information Sciences. Whereas white women make up 13 percent of the computer programmers in the US, Latinas and Black women make up 2 and 1 percent, respectively. Among software developers, white women account for 9 percent while their Latina and Black counterparts account for 1 percent each.

According to Black Girls Code, "Girls of color can learn computer science and coding principles in the company of other girls like themselves and with mentorship from women they can see themselves becoming." Black Girls Code offers girls opportunities to engage computer science in afterschool programs, weekend workshops, and summer camps. The nonprofit reflects a wider effort among educators, social entrepreneurs, and activists to cultivate greater competence and confidence in building computer-based programs and applications among populations historically underrepresented in the tech industry.

Significantly, Bryant's idea was to start with girls at the ages when ideas about one's future self begin to shape. During the transition from elementary to middle school, students continue forming their academic orientations and social identities. This is the time when girls are more likely than boys to lose interests in math and science. All students get tracked early in the education process, that is, sorted into ability groups that essentially direct their academic trajectory throughout their schooling. Students from Black, Latino, poor, and immigrant households are much more likely than their white, Asian, and affluent counterparts to be placed on vocational rather than academic tracks. These students almost never get exposure to computer science.

Further, it is not uncommon for computer programmers to begin developing an interest in programming at a young age through informal and interest-driven learning rather than formal and academic-driven learning. For example, many of the young game developers we met who write code began learning through informal networks such as online forums, affinity groups, and fan communities. They were able to explore their interest in computers in low-risk peer-driven environments, which allowed them to begin developing skills—inquiry, persistence through failure, and a growth mind-set—that researchers believe are essential in STEM fields like computer science.

THE SNOWBALL EFFECT: THE BENEFITS OF GIRLS SEEING WOMEN IN STEM

One of the more puzzling developments in postsecondary education is the decline in the percentage of women earning computer science degrees. The data suggests that colleges were actually doing a better job of producing women with degrees in computer science back in the 1980s than they are today. In 1985 women earned 37 percent of the computer science degrees awarded in the US. Since 2010 that figure has hovered around 18 percent, a significant decline. Not surprisingly, researchers have been exploring why women perform so much better in other academic fields than they do in computer science.

A 2017 study funded by Microsoft and authored by Shalini Kesar, associate professor of computer science and information systems at Southern Utah University, looks at this issue specifically. The Microsoft study surveyed 6,009 girls and women ages ten to thirty and examined their attitudes toward STEM, school, and the workplace pipeline. Similar to other investigations, Kesar's study identifies some of the factors that drive down the number of women earning computer science degrees, including, for example, lack of role models, a poor articulation by teachers of the many ways computer programing can be applied in the real world, and the underdevelopment of what researchers call a "growth mind-set" among girls in STEM.

Kesar's survey finds that female role models in STEM have a powerful effect on girls. Not surprisingly, girls rarely see women as computer science professionals or teachers, thus making it much more difficult for them to see themselves in those roles. The survey found that girls who personally know someone in STEM are much more efficacious when it comes to STEM. For instance, girls who know a woman in a STEM profession are much more likely than girls who do not to feel empowered when they are involved in a STEM activity, 61 percent to 44 percent. Also, girls who know a woman in a STEM career are much more likely than those who do not, 74 percent to 51 percent, to say they know how to pursue a career in STEM. In other words, knowing a woman in STEM can have a substantial impact on young girls' sense of ability and opportunity regarding a field like computer science. It's called the "snowball effect." The more women working in STEM, the more likely girls can see them as role models who they can aspire to emulate. The Black Girls Code after-school programs, weekend workshops, and summer camps expose Latina and Black girls to people in tech who look like them.

Girls not only lack female STEM role models in schools and other educational settings, they also lack them in the spaces where their every-day lives intersect with pop culture. Young people's relationship to pop culture is extraordinarily complex; it extends well beyond the consumption of music, fashion, television, film, and social media. Pop culture is where young people negotiate their relationships with peers and their place in the wider world around them. The media that teens engage with also reflects and shapes their sense of who they are and, importantly, who they aspire to be. As the debate sharpens about the lack of women in fields like com-puter science, observers are paying more attention to the ways in which the media environment contributes to the low number of women in fields like computer science.

Throughout the years, television and film have seldom portrayed African Americans, Latinos, and women as tech gurus and professionals, that is, as agents who can leverage technology to assert their vision of the world. From James Bond's Q to Marvel's Tony Stark, the tech makers, designers, and in-novators in pop culture are overwhelmingly white men. Popular television shows like *The Big Bang Theory* and HBO's *Silicon Valley* reiterate the narra-tive that the people leading and building the future of tech are all white men.

When the movie *Hidden Figures* (2016) was released, the positive re-sponse from the African American community was immense, in part, be-cause this type of representation in pop culture is so rare. All around the US, Black celebrities and community leaders bought out theaters so that Black students could see the film that told the story of the three Black women—Katherine Goble, Mary Jackson, and Dorothy Vaughan—who helped de-velop the math that made America's 1962 historic rocket launch possible.

And then there is the Marvel blockbuster *Black Panther* (2018). Most of the media attention focused on the box office success of *Black Panther*, which generated more than $1.3 billion worldwide. The film's financial success shattered Hollywood's long-standing assertion that movies featuring Black characters and story lines do not perform well in the international market. But many cultural commentators highlighted something else unusual about the movie. A vibrant conversation emerged around the character Shuri, the leader of the Wakanda Design Group. Audiences see Shuri doing things we rarely see young Black women doing in pop culture: using tech to solve problems, design inventive gadgets, and build the future.

Walt Hickey, culture writer at *FiveThirtyEight*, argues that the success of *Black Panther* is important but that it is Shuri, specifically, "who could change the world." He writes, "The volume of evidence shows that when audiences see onscreen representations of themselves, particularly aspirational ones, that experience can fundamentally change how they perceive their own place in the world. . . . Shuri provides a science-y role model for Black women, a group distinctly underrepresented in STEM fields."

MAKING COMPUTER SCIENCE MORE RELEVANT TO GIRLS

Many researchers believe that girls and young women struggle to identify with STEM fields like computer science. Often they do not see how the field connects to their own interests and aspirations. In Shalini Kesar's study at Southern Utah University, 91 percent of girls in grades five through twelve describe themselves as creative. Seven in ten, 72 percent, said that "having a job that helps the world is important to me." What girls say about themselves, "I'm creative" or "I want a job that makes the world a better place," does not immediately connect in their minds to computer science. In fact, computer science strikes many girls as the antithesis to being creative or impacting the world in a meaningful way.

Kesar's study maintains that STEM clubs and activities like Black Girls Code can offer girls hands-on practical experience with STEM. Through these activities, computer science becomes less abstract and more practical. In this kind of environment, girls are much more likely to experience the different ways in which computer science skills are relevant to them and their vision of the world. Importantly, computer science becomes more than just a set of technical skills; it also becomes a platform for creative expression, identity formation, and, in some instances, a catalyst for becoming an agent of social change. Learning technical skills grows girls' computer science literacy, while learning how to use tech creatively grows their computer science efficacy. As a result of STEM clubs like Black Girls Code, girls begin to develop basic computer science literacy while also gaining a greater understanding of how computer science can be applied in novel, relevant, and socially impactful ways.

Since the launch of Black Girls Code, participants have applied what they have learned by participating in everything from mobile app and game development to issues that resonate in their communities. For example, girls have designed apps that help them grapple with issues related to body image, the environment, and navigating food deserts. Black Girls Code also sponsored an event with an organization that provides support to families affected by domestic violence. During this event, participants were invited to develop apps that empower girls and their family members to recognize and deal effectively with abusive relationships. Creative opportunities like these encourage girls to see technology as a tool that empowers them to solve problems and build better communities.

The results from Kesar's study are promising. She reports that girls in grades five through twelve who were in a STEM club or activity were much more likely than girls who were not to feel empowered doing STEM (77 to 34 percent), know how to pursue a career in STEM (77 to 50 percent), and understand the jobs that are possible through STEM (75 to 53 percent). Moreover, the study suggests that STEM clubs and activities correlate with the likelihood that a girl will pursue computer science later in her education (74 percent), compared to a girl who hasn't had access to such programs (48 percent).

STEM clubs like Black Girls Code serve an important role in girls' learning environment. As we've shown, most Black and Latino youth do not have the opportunity to learn computer science in their schools, a major factor in the diversity crisis in tech. We've also learned that it is in extracurricular spaces that students feel empowered to experiment, collaborate with peers, and think outside the box in ways that schools rarely allow.

Most kids who begin to develop proficiency in computer programming at early ages likely do so in settings apart from school. For example, they may have access to informal computer science education, parents who are proficient in technology, or peer relationships in which learning how to mess around with computers and experiment with programming is a norm. In other words, they have access to informal learning environments that introduce them to the idea that computers are not only a technology for watching your favorite videos or playing games—consumption—but that computers are also for expressing ideas and building things—creation. Some researchers refer to these informal or out-of-school learning resources as enrichment opportunities.

On the surface, enrichment opportunities seem casual, even inconsequential. But social scientists believe that—like play—they are quite consequential. Studies show that students who have access to robust out-of-school learning opportunities accrue academic advantages over those who do not. In other words, student participation in a summer coding camp, an overseas trip, or an orchestra yields benefits in and out of the classroom. This may explain the gap that has widened in recent decades between what parents with college degrees spend on their children's enrichment activities compared to parents with no college. This is called the "enrichment opportunity gap." Experts believe that college-educated parents are spending more money and time on their children's enrichment activities for one reason: to better compete for admission to select universities. Over the years we have joked about "soccer moms," but it turns out they know what they are doing—giving their kids a head start in the game of life and the intense competition to get into well-regarded colleges. A study published by two economists from the University of California at San Diego calls this phenomenon "the rug rat race."

Social enterprises like Black Girls Code not only strive to close the gap in computer science literacy; they also strive to close the enrichment opportunity gap by offering a greater diversity of young people access to out-of-school learning opportunities, a space that is growing more consequential in our hyper-competitive and knowledge-driven economy.

THE GROWTH MIND-SET GAP IN STEM

Studies strongly suggest that starting young girls early in computer science clubs has long-term benefits. The sooner girls are exposed to computer science, the more likely they are to develop and maintain an interest in the field as they get older. According to Shalini Kesar's study, in middle school 31 percent of girls believe that jobs requiring coding and programming are "not for them." By high school and college that percentage increases to 40 and 58 percent, respectively.

Girls' inhibitions about STEM are influenced by a culture that socializes them to view subjects like computer science as something they cannot master. Researchers believe that girls are much less likely than boys to develop a growth mind-set in fields like computer science. A person with a growth mind-set believes that through trial and error, hard work, persistence, and inquiry they can develop greater competency and mastery over time. Girls are much more likely than boys to feel that they do not have the ability to persist and thrive in challenging STEM contexts, thus raising the stakes even higher for the potential impact STEM clubs and activities can have in the lives of girls.

It turns out that the very design of modern education—based on repetition, testing, and prioritizing knowing the answers over asking questions—may function in ways that deter girls from developing a growth mind-set and pursuing fields like computer science. Whereas a computer science course in school likely focuses on things like test scores, correct answers, and competitive learning, the computer science club likely emphasizes asking questions, taking risks, and collaborative learning.

Because STEM clubs are free from many of the bureaucratic and political constraints that make it virtually impossible for schools to experiment with new approaches to learning, programs like Black Girls Code are a great testing ground to better understand how environments that emphasize inquiry over memory, asking questions over knowing the answers, or risk taking over test taking can empower girls and other underrepresented groups in challenging subjects. Creating such an environment, many researchers believe, is key to strengthening girls' ability to thrive in computer science and building a more diverse tech talent pipeline.

EDUCATING THE AMERICA OF TOMORROW

The efforts of organizations like Black Girls Code to transform the learning worlds of students from underrepresented groups in tech are a timely response to historic shifts in the US population. The older end of the cohort continuum, age sixty-five and older, reflects, demographically, the America of yesterday. It is majority white, 77 percent, representing what demographers call a "majority-majority" cohort. The younger end of the cohort continuum reflects, demographically, the America of tomorrow. It is a population that is expected to be about 51 percent Latinos, Asians, biracial, and African Americans by the end of 2018. Demographers call this a "majority-minority" population.

The rise of a majority-minority nation has a number of implications for our society. Expert demographer William Frey of the Brookings Institution writes, "Minorities will be the source of all of the growth in the nation's youth and working age population, most of the growth in its voters, and much of the growth in its consumers and tax base as far into the future as we can see." Today, most of the major metropolitan public school systems in the US are majority-minority, which means that a high percentage of the students populating our public schools are Latino, Asian, biracial, and African American. Historically, Latino and African American students have been the least likely to have access to high-quality learning opportunities in our schools. Now that these students make up the majority of the future citizens and employees of this nation, the costs of undereducating them are simply no longer sustainable in a society and an economy that will continue to develop a demand for nuanced thinking in the STEM field and beyond.

CHAPTER 9

HACKING WHILE BLACK

Why Design Thinking Is Good for the 'Hood

In the late 1990s the US Department of Commerce launched the first series of studies examining the household adoption of the internet, titled *Falling Through the Net*. The use of personal computers and the internet was accelerating. The internet, once a technology used mainly by personnel in the military and then by universities, was coming into the homes and everyday lives of millions of Americans. The studies by the Department of Commerce warned of the new disparities that were forming around the adoption of computers and the internet. Those early studies showed, for example, that Blacks, Latinos, lower-income, and less educated Americans were being left behind in the march into the digital future. Researchers and policy makers came up with a term to describe the tech-related disparities: the "digital divide." This was a reference to a world made up of the "technology rich" and the "technology poor." African Americans and Latinos were generally regarded as being on the wrong side of the digital divide—the technology poor side.

The main factor widening the divide was economic inequality. Back then, computers were expensive, making them cost prohibitive for a majority of poor and working-class families. Compared to whites and higher-educated households, Blacks, Latinos, and lower-educated households made up a disproportionate share of the poor and the working class. Technologies like computers and the internet were luxury goods that poor households simply could not afford.

The studies included in *Falling Through the Net* were the spark behind new policy shifts that influenced everything from the federal government to local school districts. Because the studies placed a strong emphasis on the lack of access to computers and the internet, the digital divide was largely defined as a problem of access. This was the primary reason why, beginning with the last presidential administration of the twentieth century, Bill Clinton's, the federal government's strategy to bridge the digital divide was preoccupied with providing children from poor households greater access to computers and the internet. Schools and libraries became the front lines in the effort to close the digital divide. If parents in lower-income households could not afford computers and the internet, then schools, so the thinking went, should be a point of access. The strategy worked.

In 1999, 74 percent of schools with predominantly white student populations provided internet access in instructional classrooms, compared with 43 percent of schools populated by Black and Latino students. By 2005 a Black ninth-grade student was just as likely as a white ninth-grade student to attend a school that offered access to computers and the internet. Public libraries and what were called community technology centers were also equipped with computers and wired for internet access. Bridging the technology "access gap" became a high-stakes industry as billions of dollars from the federal government all the way down to local school districts were funneled into schools, libraries, and education-based nonprofit organizations with the hopes of building a more equitable digital future.

Meanwhile, some researchers and technology activists argued that the singular focus on access failed to address a multifaceted problem. They pointed out that while the access gap was certainly important, so was what they called the "skills gap." Over the years, countless studies have documented the disparities that exist in the skills and social networks that support how different populations use the internet. What these studies consistently conclude is that simply providing people access to the internet is not a sufficient measure of digital equality. Rather, as the technology has become more dominant, it has become vital that people develop the ability to use the internet in what scholars call "capital-enhancing" ways. This includes, for example, being able to use the internet to improve your social networks, employment prospects, or participation in civic life.

The view is widely shared today that the solution to the digital divide is more complicated than simply providing access to technology. In fact, there is widespread recognition that the disparities in tech include multiple gaps, including the access gap, the skills gap, and the participation gap. But schools are still struggling to reorganize their mission to build a more equitable future around this multidimensional definition of the digital divide.

To be fair, addressing the access gap is much easier than addressing the skills gap. Simply by shifting portions of their budget to acquire more technology—smart boards, desktops, laptops, tablets, and software—schools can close the access gap. But closing the skills gap requires designing whole new curricula, developing an army of trained teachers, and expanding how we think about digital literacy in a world of constant connectivity, artificial intelligence, big data, and "fake news."

In my own research I have observed how schools operate from the assumption that the mere acquisition of technology is an indicator of better learning futures. But more technology is not the solution to building a more equitable society. Rather than investing in more technology, schools should be investing in curriculum and instructional expertise that builds up young people's ability to use technology in novel and capital-enhancing ways. Rather than putting their faith in new tech gadgets, schools should be betting on the creation of classrooms that function more like innovation labs rather than way stations for more standardized testing. It is time for schools to see technology as a tool for supporting richer learning environments and not an indicator of richer learning. In a series of reports for the *New York Times* in 2018, Nellie Bowles explains how tech-literate parents in Silicon Valley are opting for school and home environments that actually limit their children's access to screen technologies. And while their decision to limit their children's exposure to screens is a mark of privilege, it is also a reminder that in our knowledge economy, learning to use creative-thinking and problem-solving skills is just as important as learning to use technology.

FUTURE-READY

As the presence of technology has become ubiquitous in our lives, in our economy, and in our schools, there is a tendency to view tech skills such as coding, software development, and web development as the dominant human capital asset in the world today. But that is not entirely true. Instead, the most important tech-related human-capital asset is the ability to use technology in novel, inventive, and impactful ways. Thinking skills, not tech skills, are what really drive our knowledge economy. The ability to generate and bring new and timely ideas into the world has become a preeminent skill in the knowledge economy. Moreover, the ability to think in nuanced ways about the implications of artificial intelligence for work, privacy, and society is growing more urgent. US public schools should be built to help young people develop the skills that are a true fit for a society and an economy undergoing a profound transformation.

The state of thinking that governs public education is strikingly short-term. In my many conversations with good-intentioned educators, there has been a remarkably consistent emphasis on students being "college-ready" or "career-ready." A look at the education data in the US clearly indicates that most students, especially Black, Latino, and lower-income students, are not college-ready. For example, in 2015 about 35 percent and 37 percent, respectively, of Black and Latino eighteen- to twenty-four-year-olds were enrolled in college. By contrast, 63 percent of Asian and 42 percent of whites the same age were enrolled in college. Black and Latino college students are less likely than their Asian and white counterparts to leave college with a degree in hand. Asian students, at 71 percent, are the most likely of all students to graduate from a four-year college within six years, while Black students, at 41 percent, are the least likely. Both white students, at 63 percent, and Latino students, at 54 percent, while less likely than their Asian counterparts, are more likely to earn a degree in six years than Blacks.

The "career-ready" mantra is problematic too. The emphasis on career-readiness is especially concerning when you consider that young people entering the economy will likely never experience a career in the way that previous generations have. First, most young people will work an average of twelve to fifteen jobs in their lifetime, suggesting that they will constantly be moving from one job to the next. Second, most companies are moving away from the long-term employment or career model, preferring instead to hire contract employees, temporary workers, or gig laborers. In short, the career-readiness mantra is anachronistic and out of step with the world of work that most young people will experience. What should schools be doing? Instead of preparing students to be college-ready or career-ready, schools must start producing students who are what I call "future-ready."

The skills associated with future readiness are geared toward the long-term and oriented toward navigating a world marked by diversity, uncertainty, and complexity. Career-ready skills encourage students to focus on getting a job today. By contrast, future-ready skills encourage students to be responsive to an economy that will continue to evolve and demand skills that are not even discernible today. Furthermore, future-ready skills involve leveraging technology to intervene in and build the world in ways that are both original and valuable. Whereas the career-ready approach, at best, prepares students for the world that we inhabit today, a future-ready approach prepares students for the world we will build tomorrow.

In their book *The New Division of Labor: How Computers Are Creating the Next Job Market*, economists Frank Levy and Richard J. Murnane contend that the introduction of computers into the workforce fundamentally changed the occupational structure in the US. Levy and Murnane, like many other economists, maintain that knowledge-driven economies privilege workers who are highly educated and highly skilled. These workers are more likely to be employed than their less educated and less skilled counterparts. They are also more likely to command jobs that pay well. Economists call this "skill-biased technical change."

In a skill-biased economy driven by computers, Levy and Murnane argue, there are basically two kinds of jobs—those that involve skills that complement computers and those that involve skills that computers can replace. In the former category, individuals have skills and expertise that cannot be easily automated or performed by smart machines, such as a novelist or curriculum designer. Individuals in the latter category have skills that can be easily replicated by smart machines, such as a worker who accepts and tracks bank receipts. Levy and Murnane contend that schools should be cultivating a repertoire of skills that are difficult for smart machines to perform by themselves, such as solving uncharted problems, writing a novel, or working with and applying new information in inventive ways.

Levy and Murnane first issued their economic analysis in 2000. Nearly two decades later the arrival of smart machines, robots, and artificial intelligence are accelerating the pace of change and raising serious questions about the future of work. Today the major concern is whether or not intelligent machines will render most human laborers obsolete in the so-called jobless future.

A 2017 report by the McKinsey Global Institute finds that as early as 2030 about one-third of the American workforce may have to find new forms of employment due to automation. These changes, the report asserts, "imply substantial workplace transformations and changes for all workers." Assessing historical trends, McKinsey adds that roughly 9 percent of labor demand in 2030 will be in occupations that have not existed before. Some of these new occupations will almost certainly be related to helping smart technologies perform tasks that address human needs in a rapidly evolving society. One of the big challenges that schools face, of course, is preparing young people for meaningful work in a world we cannot yet imagine. In truth, this scenario—attempting to prepare students for a society and an economy we cannot see—is not unprecedented. For example, twenty years ago schools were not preparing students for a society and an economy in which tech advances like smartphones, artificial intelligence, and autonomous vehicles were either fully integrated into everyday life or just on the horizon.

Building on the ideas of Levy and Murnane, MIT economists Erik Brynjolfsson and Andrew McAfee posit that the human ability to ask novel questions will remain highly valuable even in a period characterized by rapid computerization and automation. Brynjolfsson and McAfee contend that ideation skills—the ability to think outside the box about complex problems for which there are no routine solutions—is a human skill that will continue to have value in the economy of tomorrow. Computers may be powerful tools for organizing people around the world for a social cause, such as Black Lives Matter, but they are not very good at knowing that they can be used this way. Humans, Brynjolfsson and McAfee maintain, are much more likely to ask "What if?" or "How can we?" People who are good at ideation will not be easily replaced by smart technologies for one main reason: they will be able to create new tasks and jobs for humans to perform. Rather than face a jobless future, they will build a creative future.

Students who are future-ready have the intellectual ability to grapple with novelty and complexity and also to see opportunity where others do not. Future-ready skills involve nuance, expertise, and inventiveness. In my own work I have developed an interest in one future-ready skill set: design thinking.

"DESIGN IS TOO IMPORTANT TO BE LEFT TO DESIGNERS"

There are many ways to describe design thinking, but it generally involves the ability to respond to specific challenges and solve problems in creative ways. I visited many design studios to get an up-close view of what makes design such a critical future-ready skill. The studios feature open spaces for interdisciplinary conversations and collaboration, deep engagement and problem solving; plenty of whiteboards to visualize the ideas that team members develop; and work spaces where team members make their ideas tangible—as a sketch, a prototype, or a digital artifact.

When I visited IBM-Design, they were going through a massive internal reorganization, hiring young creatives from a variety of disciplines—the arts, social science, and engineering—to help push the company's thinking into the future. I also visited the LUMA Institute in Pittsburgh. Once known for its role in steel and in the manufacturing economy, Pittsburgh has become a hub for innovation led by a variety of efforts by professors from universities like Carnegie Mellon and spurred by the presence of tech companies—including Google, Uber, and artificial intelligence start-ups like ARGO AI, just to name a few—that are launching new initiatives.

During my visit to the LUMA Institute, I sat down with the director, Chris Pacione. He had written a piece for the design magazine *Interactions* that intrigued me. In that article Pacione makes an excellent case for design as a core literacy, something that all students should have a chance to learn. Design, he argues, should be as pervasive as reading, writing, and arithmetic. Pacione lays out the case for how design literacy, or "pervasive competency in the collaborative and iterative skills of 'looking' and 'making' to understand and advance our world," could represent a breakthrough in the history of common literacy. He is among a group of design professionals who are making the case that "design is too important to be left to designers." The skills that design fosters—inquiry, ideation, experimentation, prototyping, and iteration—are some of the most valuable skills in today's knowledge-driven economy. They are, quite simply, skills that humans perform much better than machines. When designers see the world, they do not see the world as it is, but rather as it could be. Design, we could say, is the process designers execute to make the world they imagine real.

At LUMA they emphasize the skills associated with "looking" and "making." Looking is a set of techniques that involves observation, research, inquiry, and insight-building. Making is a set of techniques that involves crafting, building, prototyping, and mobilizing ideas into tangible form.

Design thinking in the context of the business world is used to create and maintain a strategic advantage in an economy that increasingly sells knowledge and expertise as products or services. Critics maintain that design thinking is really a clever form of branding, a way for American businesses to reestablish their dominance in a more competitive global economy. Design thinking, from the view of critics, is packaged and co-opted by business consultants and university programs to profit from the next new thing in enterprise capitalism.

The concept of design thinking has become trendy, which raises legitimate questions about the validity of claims that evangelists make about the practice. However, the core of design thinking—creative problem-solving—is not a new form of human behavior. Indeed, the components of design thinking, such as inquiry (call it research), ideation (call it brainstorming), or prototyping (call it making something out of nothing), have been around for centuries.

My interest is not so much in the claims that design-thinking gurus make to give businesses a competitive position. Instead, my interest focuses on what I like to call the "design disposition." This is the ability to, first, see the world not as it is but rather as it could be and, second, to figure out how to make that vision real in some way. Design thinking is as much about human efficacy as it is corporate efficiency. In the new innovation economy, the elements of design thinking are being deployed by young creatives to address the challenges associated with social and economic inequality.

Design thinking is catching on in some schools but only those that have the resources to invest in the curriculum, instruction, and professional development that expose students to the skills associated with the practice. Most schools, however, do not have sufficient resources to invest in innovative materials and teacher training. Moreover, most schools are not configured to encourage students to see themselves as agents of change in their school, community, or society. Students are not socialized to conceive of themselves as designers, that is, as people who see the world not as it is, but rather as it could be.

How do we, then, teach a greater diversity of young people future-ready skills? Some of the young creatives that I met during my research have made questions like this one the centerpiece of their social and entrepreneurial mission. They recognize that traditional education must be disrupted in order to make schools and learning relevant to the workers and citizens of the future. Further, they recognize that the future of learning is more about what students do with technology rather than simply providing access to more technology.

Our current public education system is severely underequipped to prepare the nation's most diverse student population in history for a society and economy increasingly shaped by smart technologies and a demand for higher-order cognitive skills. Among the skills that schools should be helping students cultivate, none may be more important than the ability to think outside the box to generate and transform insight into tangible solutions.

HACKATHONS: JOB FAIRS FOR THE TWENTY-FIRST CENTURY

One of the most creative ways design thinking has infiltrated the world outside the design professions is through hackathons, those carefully coordinated events that bring people together to invent things fast. Hackathons cram all of the components of design thinking—inquiry, brainstorming, experimentation, prototyping, and iteration—into a two- to three-day sprint. For years now, hackathons have been a fountain of innovation in the tech world. Facebook is one of the companies that recognizes the crucial role hackathons can play in sparking innovation. The social media giant, in retrospect, was likely the product of late-night coding and rapid prototyping sessions—hackathons—that focused intensely on building out the basic architecture of the platform. For years the company championed the "break things fast" ethos in which it had been born. That approach to innovation, of course, has produced some significant problems for Facebook and the millions of people who use the social platform.

A Facebook engineer explains the influence of hackathons at Facebook this way: "Hackathons are a chance for engineers, and anyone else in the company, to transform the spark of an idea into a working prototype and get other people excited about its potential. We're a culture of builders, and hackathons are our time to take any idea—big or small, sane or crazy—and build it into something real for people to react to." Facebook realized early on that giving employees the freedom to create and build new features is a way to keep the platform fresh and evolving as it becomes more massive in scale and corporate in its operations. Some of the more popular features on Facebook are the result of these all-night hackathons that have become a source of creativity and camaraderie for Facebook employees. Facebook Video and Timeline are the result of hackathons. It was an internal hackathon at Facebook that led to the creation of one of its most widely used features, the "Like" button.

Hackathons have spread beyond tech companies like Facebook. Regarding the popularity of hackathons on college campuses, Steven Leckart writes in the *New York Times* that these events "encourage students to tinker with new software and hardware and challenge themselves." Hackathons, the article suggests, have become a focal point among tech and media companies, which now sponsor college hackathons as a way to generate new ideas for their companies and to recruit new talent— like job fairs for the twenty-first century. A partner from the venture capital firm Kleiner Perkins Caufield & Byers proclaims, "The best talent are at these opt-in courses called hackathons. . . . If you're not at a hackathon, you're at a disadvantage."

Throughout most of their history, hackathons have largely been coordinated by and for the tech bros who have become the face of our tech and knowledge-driven economy. But the face of the hackathon is changing as a new generation of community leaders, educators, and tech activists, well, hack hackathons by transforming them into community-based design challenges that articulate a new vision for what hackers can create. Community-driven hackathons are important for two reasons. First, they invite a more diverse population to mobilize and build tech solutions. Second, the tech solutions that community-driven hackathons inspire are often influenced by ideas, expertise, and experiences that are generally excluded from the culture of hacking and innovation in the tech world.

In Oakland, educator and social entrepreneur Kilamah Priforce founded the Qeyno Group (formerly Queyno Labs) and started using hackathons in 2013 to empower, in his words, "high-potential youth in low-opportunity settings." Today, Priforce's Hackathon Academy runs three-day "pop-up schools"—hackathons during which teens use tech to solve real-world problems that affect them and their communities. "Hackathons have been around for a while in Silicon Valley and Ivy League institutions, but if you walk into your average hackathon they are usually too white, too male, and too unhealthy," Priforce says. He views hackathons as a way to introduce a more diverse community of young people to the principles of design and tech-inspired social innovation.

Priforce named the Qeyno Group after his brother, who was tragically shot to death as a teenager. Priforce and his brother came up the hardscrabble way. Their home life as kids was so unsettling that they were forced to move into a group home in Brooklyn. Needless to say, life was challenging for the brothers. In the group home Priforce became a voracious reader. He read all of the books in the home's limited library at least three times each. When he asked to go to the public library so that he could find more books to read, he was told no. Priforce and the other kids were not allowed to visit the city's libraries or museums. Frustrated with the situation, Priforce decided to take action.

"When I was eight years old . . . I had a three-day hunger strike against my group home, in order to get more books," he says. The hunger strike earned him and the other kids the right to go to the library. Years later, he likened the group home to a prison, but the experience lit a fire in him for literacy, empowerment, and social justice that continues burning today. The Qeyno Group and its Hackathon Academy are the result of that fire.

INCLUSIVE INNOVATION

The hackathons organized by the Qeyno Group bring elements of design thinking into the neighborhoods where Black, Latino, and underserved teens live. Priforce says that the tragic killing of Trayvon Martin, the unarmed Black teen who was gunned down by neighborhood watchman George Zimmerman in 2012, inspired the idea for the hackathons. Martin's death reignited a long-running debate about the use of deadly force by police in Black communities. Priforce asked himself, "Could an app have saved Martin's life?" The question was the catalyst for a greater challenge: to begin cultivating a generation of young people who develop and use tech for social justice and social impact.

When Priforce moved from the East Coast to Oakland, he visited the schools and noticed what he considered a lack of leadership. Nearby, San Francisco and San Jose were emerging as powerful tech and innovation hubs, but Oakland, which at the time was much more racially and ethnically diverse, might as well have been in another country. Priforce decided that if Oakland schools were not going to provide Black and Latino youth with opportunities to develop future-ready skills, he would.

His first event was billed as a Black Male Achievement hackathon. It was part of the city of Oakland's first Startup Weekend. A growing number of cities are hosting events like Startup Weekend to bring together and support aspiring entrepreneurs. One of Priforce's main goals of his first Qeyno hackathon was to introduce students to coding, art design, and product development. This event established the template for future iterations of the Qeyno hackathon: First, create an environment that challenges high-ability low-opportunity youth to see themselves as agents of change. Second, equip them with the social, human, and digital capital to spark their belief that change is possible. Third, build a curriculum that allows young people to get hands-on experience generating and implementing solutions to real problems. Fourth, enlist tech professionals, designers, artists, and educators to volunteer their time and expertise to work side by side with students to turn their idea for an app or an intelligent solution into a prototype. And finally, begin creating a tech innovation ecosystem that fosters diversity and inclusion. Priforce calls the Hackathon Academy model "inclusive innovation."

Since that first event in Oakland, Qeyno has conducted hackathons for youth in New Orleans, Detroit, Philadelphia, and for Native American youth in New Mexico. Thematically, Qeyno's hackathons typically challenge participants to design solutions for problems in public health, education, financial inclusion, social justice, and sustainability. On at least one occasion, Priforce pushed the envelope a bit too far, according to some of his critics. After the death of Michael Brown in Ferguson, Missouri, CNN commentator and former Bay Area activist Van Jones phoned Priforce and asked him about the prospects for bringing police and young Black Ferguson residents together to brainstorm ideas to foster better relations. The idea that Black youth would actually sit down and design solutions with law enforcement struck some in the activist community as too conciliatory. Ferguson activists even threatened to Molotov the event.

After some initial planning, the Ferguson event never happened. However, when Oakland experienced its own conflict between police and local Black residents, Qeyno arranged a hackathon with Black teens and police officers in that city. Oakland activists protested the idea, but Qeyno went forward anyway, convinced that any attempt to resolve the tension between Black communities and the police required the two groups working together.

CULTIVATING A DIVERSE GENERATION OF TECH TALENT

The core purpose of the Qeyno hackathon is to empower young people to begin using technology in ways that enhance their capital. By learning how to design mobile apps and other tech solutions, for example, young participants grow their human capital. Because they get to work with technology and design professionals, these teens also grow their social capital. The opportunity to design community-oriented solutions grows their sense of civic engagement and political capital too.

The variety of apps that teams attempt to build during hackathons illustrates the ways these problem-solving sessions can benefit the community at large. Participants have designed apps that encourage their peers to make healthy food choices or read more books. One team designed an app for Black students who attend predominantly white private schools. The app offered Black private school students an opportunity to build an online community to grapple with feelings of social isolation. One app promotes young women's self-esteem while another funds computers for low-income youth. Solutions such as these do not simply promote technical achievement; they also promote civic engagement.

Each hackathon requires meticulous coordination and the courting of sponsors and tech professionals. The high volume of teens seeking admission is a testament to the value of enrichment activities like the Qeyno hackathons. The Philadelphia hackathon resembled many of the design studios that I had visited. The physical space was open, inviting participants with diverse backgrounds and expertise to share ideas, collaborate, and rapidly prototype together. There was plenty of space in which to visualize, refine, and iterate ideas. The teens worked closely with designers, artists, and programmers to translate their ideas into tangible products. But more than anything it was a space for teens to take on roles that schools rarely, if ever, encourage. The teens were leaders, artists, designers, active learners, and innovators. Moreover, they asked, "What if?" and "How can we?" to generate ideas, sketch their mobile applications, and storyboard scenarios that proposed solutions to the community problems they identified. In short, they were practicing the design skills common in the tech and innovation economy.

Hackathons have created an environment in which young, Black, creative, civic, and academic talent can flourish. Moreover, the Qeyno hackathons suggest that while university athletic programs excel at identifying and recruiting Black athletic talent, Black tech and academic talent have largely gone unnoticed and untapped. Take George Hofstetter, a young tech enthusiast and one of the first participants in a Qeyno hackathon.

Hofstetter was fourteen when he participated in his first hackathon. He was one of the few Black students enrolled in his predominantly white private school, and Oakland-area enrichment opportunities like Brothers Code and the Hackathon Academy invited him to explore tech, design, and coding with peers who looked like him and identified with some of his struggles in the world. "To actually know that there are other African Americans who like to do what I do was so inspiring," Hofstetter told a Bay Area journalist.

During the Black Male Achievement hackathon, Hofstetter pitched an app called CopStop. For many of the young African American participants, including Hofstetter, the hackathon was an opportunity to deploy a tech solution to a local problem that affected them personally. CopStop was a digital facsimile of what has become known in the African American community simply as "the talk."

In the aftermath of the many high-profile police killings of unarmed African Americans, "the talk" represents the conversations that many Black parents are having with their kids about how to be safe when interacting with police. Among other things, "the talk" recognizes the vulnerability of Black lives in the world today and facilitates a conversation about specific techniques that Black teens should exercise when confronted by law enforcement. Some of the techniques include keeping your hands visible or asking for permission to reach into your pocket or the glove compartment of a car. CopStop is a mobile version of "the talk." More precisely, it is a smart mobile intervention that shares tips that some believe have the potential to save young lives. Among other things, the app alerts people in your contacts list if you are stopped by a police officer and allows the users to video record the encounter.

The app can also notify someone, a person of the user's choosing, about his or her location when encountering police. CopStop reminds users of their rights when they are approached by law enforcement. In 2016 Hofstetter told Yahoo! News that he created the app to "help eliminate the feelings of danger, anxiety, and inequality brought on when interacting with law enforcement." After refining the app and updating some of the coding, Hofstetter released CopStop in the Apple App Store in January 2018.

Zachary Dorcinville, a young African American teen from Orange County, New York, also found the hackathons personally reaffirming and inspiring. When Dorcinville discovered the Qeyno hackathon in New Orleans for Black teens he emailed Priforce to express his interest. Dorcinville was ecstatic when he received an invitation to participate, but there was one problem. He lived in New York and the event was in Louisiana. Qeyno provided hotel accommodations and meals, but participants had to supply their own transportation. Dorcinville's family did not have the funds to support his air travel, but this did not deter him. Dorcinville created a crowdfunding campaign on GoFundMe to help raise money for the trip. He raised just enough to make it to New Orleans.

Dorcinville got a chance to attend another Qeyno hackathon, this time in Philadelphia, which was much closer to home. About sixty young people participated in the three-day event. All participants were required to pitch their ideas on Friday, the first day of the three-day event. Dorcinville pitched an app he called Recupery, which he described this way: "It provides an easy gateway for children and adolescents to gain access to a counselor via videoconference or a phone call if they don't have access to speak to a counselor in person."

Recupery was selected along with eleven others to be developed into a prototype during the hackathon. Dorcinville and his teen collaborator were partnered with a designer, a programmer, and an artist to build a prototype. When it was time to pitch Recupery to a panel of expert judges, Dorcinville told the story of John, a sixteen-year-old boy who suffers bouts of depression. In many ways, John was Dorcinville. They were both sixteen and they both struggled with depression.

The smartphone-enabled counseling sessions that Dorcinville proposed represent what tech and mental health professionals would call a digital platform for behavioral intervention. I have served as a referee on panels assembled by the National Institutes of Health to evaluate the potential impact of tech-based solutions for health problems. Dorcinville's app has features that are strikingly similar to what university professors and health and technology professionals are proposing to bring health care and client therapy into the smart future. Billions of dollars are being invested in determining the most effective ways for smartphones and artificial intelligence to provide more efficient, personal, and effective health and social services. In the environment created by the Qeyno hackathon, Dorcinville, at the age of fifteen, was already designing an application that engaged similar ideas.

Victoria Pannell is another young African American who participated in the Qeyno hackathon in Philadelphia. A fifteen-year-old student at the time of the hackathon, Pannell was inspired by the event to think about how smart technologies might address real-world problems. When Pannell was thirteen, she was chosen to play the role of a thirteen-year-old girl who was a victim of sex trafficking in a public service announcement. When she met the girl she would be portraying, her life was changed forever. Pannell dedicated herself to becoming an activist against sex trafficking. A year or so after the Philadelphia hackathon, Pannell built an app called STOP, an acronym for Sex Trafficking Operations Prevention. Appearing at a New York City event to discuss equity in STEM education, Pannell presented the STOP app. "STOP," she told the audience, "is a platform that empowers us to blow the whistle on sex trafficking in our communities." The app offers features that let witnesses report sex-trafficking activities as well as allow victims to call a hotline for help.

The minds behind each of these projects deployed design thinking as a medium to engage critical issues facing communities and people in distress. For participants like Hofstetter, Dorcinville, and Pannell, the hackathon was an invitation to use technology to realize the power of their own agency. Their stories also highlight young African Americans' changing relationship to technology.

BEYOND THE DIGITAL DIVIDE

In the years since its inception, the digital-divide narrative has become a compelling story, but one that has also limited educators', researchers', and policy makers' capacity to see and understand the changes that are happening right before their eyes. Between 2004 and 2010 African American and Latino teens' use of the internet underwent a stunning transformation. When I began talking to Black and Latino teens back in 2007, one thing was clear—they were not technology laggards. In fact, they were emerging as tech early adopters, trendsetters, and civic innovators.

Black and Latino teens' use of the internet began to accelerate largely because of mobile phones. Soon a reality that had seemed unthinkable at the start of the new millennium took form: Black and Latino teens were spending just as much time online as their white and Asian counterparts. Furthermore, they were social media "power users," a term used to describe populations whose use of technology is more frequent compared to others. But much of the focus was on how these power users used mobile phones and the internet to consume content such as games, video, music, and social media.

What is largely hidden are the ways Black and Latino teens use the internet to communicate, connect, and create content that resonates with their own sense of self and their view of the world. They understand how social media can be used to craft safe online spaces in a world in which living while Black or Latino is still perilous. Black teens were the first among all youth to adopt Twitter at scale. They are the catalyst behind the most powerful collective force in Twitter—Black Twitter. By 2010 Black and Latino teens' savvy use of social and mobile media suggested they were innovators, early adopters, and trendsetters. Black and Latino youth's engagement with technology far exceeds the limited notions held by educators and policy makers who continue to think of them as laggards, late adopters, and trend followers.

Social entrepreneurs like Priforce understand that if afforded the opportunity, Black and Latino teens can in fact use technology in transformational and capital-enhancing ways.

DESIGN FOR SOCIAL CHANGE AND SOCIAL JUSTICE

The skills teen participants of the Qeyno hackathons use so well—interdisciplinary collaboration, ideation, problem-solving, prototyping, and iteration—are the skills that are increasingly in demand in a knowledge economy. They are also the skills that empower humans to exercise agency and creativity in a world increasingly shaped by intelligent machines. Teens who participate in the Qeyno hackathons develop a disposition that is inquisitive, creative, and biased toward action. Hackathon Academy's pop-up schools encourage students to apply the skills of design not in the pursuit of money or personal benefit but rather in the pursuit of social justice and societal benefit.

The mission of the Qeyno Group represents so much of what I have discovered about the new innovation economy. It is equal parts tech ingenuity and social ingenuity. Its most nimble agents have not just adopted technology; they have applied it in ways that most Silicon Valley tech bros could never imagine.

Young creatives like Priforce are moving into the education space in American culture to do what schools have essentially been unable to do. If schools have been slow to address long-standing academic learning gaps or the demands of the knowledge economy, a growing number of innovative educators and social entrepreneurs are mobilizing their resources to create more equitable, future-ready learning environments.

CHAPTER 10

WOKE

The Rise of Connected Activism

On April 23, 2016, President Barack Obama conducted a town hall meeting with a group of young London leaders as part of a three-day trip to the UK. The president abandoned his customary suit and tie for a more relaxed ensemble that underscored his warm and friendly relationship with young people. In his opening remarks, Obama expressed his appreciation for "the chance to meet with young people and hear from them directly." He said that hearing from this demographic "gives me new ideas and . . . underscores the degree to which young people are rising up in every continent to seize the possibilities of tomorrow."

As the town hall proceeded, young attendees asked the president about a variety of topics, including Northern Ireland, the T-TIP agreement, East Africa, and leadership in a world marked by political polarization. One young woman named Maria struck a more personal note when she described her gender identity as "non-binary." She talked about how terrifying it is to live in a country in which "non-binary people . . . literally have no rights." Maria asked the president to address civil rights and LGBTQ issues for people "who fit outside the social norms." Then Louisa, a self-described "climate change campaigner," asked the president which social movements "have made you change your mind while you've been in office and inspired you to do things?"

In response to these questions, President Obama acknowledged how LGBTQ activists and his own two daughters had enlightened him toward a favorable position on same-sex marriage. He noted his role in the historic

Paris climate agreement. He pointed to the rising unrest in the US around race and the use of police force, and he named Black Lives Matter as a great example of how young activists can bring attention to the issues that they care about. The president also issued a warning about youth-driven activism: "Once you've highlighted an issue and brought it to people's attention and shined a spotlight, and elected officials or people who are in a position to start bringing about change are ready to sit down with you, then you can't just keep on yelling at them."

The President asserted that it's okay to make noise and occasionally act a little crazy to get attention and shine a spotlight on an issue. "But," he said, "once people who are in power and in a position to actually do something about it are prepared to meet and listen with you, do your homework; be prepared; present a plausible set of actions; and negotiate and be prepared to move the ball down the field even if it doesn't get all the way there."

Back in the US, reports of this meeting framed the president's comments as criticism of Black Lives Matter. A *New York Times* headline read "Obama Says Movements Like Black Lives Matter 'Can't Just Keep on Yelling.'" The *Washington Post* ran a similar headline: "Obama Counsels Black Lives Matter Activists: 'You Can't Just Keep on Yelling.'" In fact, the president's comments struck a familiar note about the Black Lives Matter movement specifically and youth activism more generally. Youth activism these days is critiqued as more style than substance, more tech savvy than political savvy, more vanity-driven than policy-driven—all obvious references to young activists' use of social media as a means to participate in civic life.

Obama was not the only high-profile critic of youth-driven social movements. Media mogul Oprah Winfrey suggested that unlike the participants in the 1960s civil rights movement, the young activists driving Black Lives Matter have no specific goals or clear asks. However, Obama and Winfrey, in their comments, underestimated a movement that was more robust than they and most others realized. Critics of Black Lives Matter have been largely unaware of the movement's evolution and the substantive impact young activists desire to make. Moreover, critiques like these fail to adequately appreciate the complex features of contemporary youth activism and the evolving models of political engagement they inspire. In fact, young activists have adopted many of the features of the new innovation economy—collaboration, inventive uses of technology, crowd power, and the mantra of failing fast and trying again—to pursue their vision of a more equitable and just society.

THE CIVIC DECLINE NARRATIVE

One of the most persistent criticisms of millennials is the claim that they are so reliant on social and mobile media that they are less involved than previous generations in the world around them. Young people's preoccupation with technology, it is strenuously argued, has undermined our sense of community and the future of democracy. According to this view, the integration of social media into daily life has made millennials less civic-oriented than their parents and grandparents. Scholars who study the quality of civic life call this the "civic decline narrative."

Critics claim that by just about every measure, young people, to paraphrase the political scientist Robert Putnam, are playing the civic game less than their older counterparts. Younger Americans, we know, vote less than older Americans. Younger Americans, critics assert, also consume less news than older Americans.

Another supposed indication of the erosion of civic engagement is the steep drop-off in affiliations with political parties. And it is true: millennials are much less likely than their older counterparts to report belonging to one of the two major political parties. In his book *Bowling Alone*, Putnam points to a decline in what he calls "grassroots activism"— political protest or mobilization that is inspired by a set of local or national conditions. Political scientists also believe that political expression—writing to an elected official, signing a petition, writing an article or letter to the editor, or making a speech—is on a downward trend.

Researchers have identified several reasons to explain young people's retreat from civic life. Some studies, for example, suggest that younger people are much less likely than their older counterparts to trust traditional sources of political influence and power. Others argue that apathy or lack of interest in public affairs is a significant factor. Additionally, millennials are frequently accused of narcissism. Historically, younger citizens tend to feel less politically efficacious than older citizens. But no factor is more resonant in the critique of young people than the claim that media and technology have contributed mightily to declining participation in civic life.

Millennials came of age in the most media-rich households in history. Over the years multiple entertainment technologies—DVDs, video games, desktop and laptop computers, smartphones, and tablets—have entered our homes. Consequently, one scholar claims, our homes have become "wired castles," a description that highlights the extent to which entertainment platforms wall us off from the outside world and diminish our interest in community involvement. Critics and social scientists alike maintain that increased attachment to screens, especially mobile devices, makes all of us—but especially children, teens, and young adults—less social, connected, and communicative. Social media, in this context, is an oxymoron precisely because it presumably encourages less "authentic" social contact.

Millennials—the most connected generation in American history—feature prominently in this version of the civic decline narrative. These circumstances have led cultural critics and researchers to conclude that millennials are too busy "tweeting," "snapping," or posting to Instagram to care about anybody or anything beyond themselves. The greatest social cost of social isolation, according to Putnam, is the loss of social capital. Nan Lin, a longtime scholar in the area of social networks, defines social capital as those resources available to individuals through social connections. When members of society invest less in one another, they not only become social-capital poor, society suffers too. We spend less time staying informed about our communities and world affairs. We spend less time participating in civic organizations and are less likely to support charitable causes and organizations. Simply put, we are less likely to care about creating a better world, less likely even to think that we have the capacity to do so.

The narrative that millennials are disengaged with civic life focuses primarily on legacy civic institutions, including political parties, community-oriented organizations like bowling leagues, and traditional forms of civic expression like writing to the local newspaper editor. Still, there is mounting evidence—anecdotal and empirical—that millennials are involved in the civic sphere but in ways that researchers have not adequately measured or understood. Thus, an interesting question emerges: What if the claim that young people are disconnected from and disinterested in the world around them is an overstatement or just plain wrong?

THE CIVIC INNOVATION NARRATIVE

In contrast to the civic decline narrative let me propose an alternative—what I call the "civic innovation narrative." Whereas the civic decline narrative asserts that there is a steep decline in political engagement, the civic innovation narrative contends that what is really happening is a remaking of political engagement. The civic innovation narrative offers a portrait of a civic culture that is creative, iterative, and fit for the networked age.

What about the claim that millennials consume less news, a key component of political knowledge and participation? In a national survey of young adults that Vicky Rideout and I conducted with the nonpartisan research organization NORC at the University of Chicago, millennials' use of smartphones to go online was decisive compared to other means. Sixty-eight percent reported going online with a smartphone, compared to 8 percent who reported going online via a desktop computer. According to the Pew Research Center, people who get their news on their mobile devices throughout the day tend to turn to more resources, get news from new sources, watch news videos, read in-depth news articles, and send and receive news through their social networks. The fact that millennials are more likely than their older counterparts to engage news this way suggests that they may actually be exposed to more news sources rather than fewer.

In the age of social media and "fake news," the concern may not be lack of access to news but rather the quality of the news we consume. Millennials are living testimony to the often uttered view that in the connected world we no longer find the news; rather, the news finds us. What we still do not know, however, is whether or not the ambient qualities of news—the constant news alerts and mobile updates—are leading to more in-depth engagement, political knowledge, and participation in civic life.

Nowhere is the formation of civic innovation more apparent than in the rise of what I call "connected activism." Broadly speaking, young people practice connected activism through informal modes of political activity, inventive uses of technology, creative political expression, and direct action against powerful institutions. Rather than submit to formal politics, read print news, associate with legacy civic organizations, or view voting as the only expression of civic engagement, millennials are devising new pathways to pursue civic investments that are responsive to our times and their vision of political engagement.

I have identified five basic components of effective connected activism: it is mobile, visual, spreadable, scalable, and impactful. First, the benefits of mobile devices in the civic sphere are notable and substantial. For example, smartphones have become a tool for real-time communication, civic media making, and political organizing. As the capabilities of smartphones expand, activists use them to capture and share photos and videos that reinvigorate the tradition of citizen journalism. A second component of connected activism is the visual nature of social media communication and the creation of media content—photos, videos, memes, graphics—that can expand and enrich storytelling and political expression. The visual aspect of connected activism speaks to the broadcasting capabilities of social media channels like Twitter, YouTube, Facebook, and Instagram and the affective power of visual content to stand as witness to social injustice and to catalyze community dissent.

A third component of connected activism is the spreadable nature of social media content, that is, the ease and speed with which media and messages can be circulated and consumed. The shifting media landscape in which we now exist makes it easy for connected activists to produce and circulate content beyond the corridors of corporate media. Social media, by design, is spreadable media—media that is created to move fluidly across computer-mediated social networks. A fourth feature of connected activism is the scale at which communication and organization can take place. Social media improves the opportunities to engage crowd power—through crowdsourcing and crowdfunding—to bolster political causes. When Black Lives Matter activists traveled to participate in local demonstrations against law enforcement's use of deadly force, they often turned to crowdsourcing to connect to local activists, coordinate local protest activities, and even find accommodations.

A final component of connected activism is the degree to which mobile, visual, spreadable, and scalable features of connected activism enhance the prospects for greater social impact. Due to the internet's ability to quickly multiply the number of participants who can connect to a movement or circulate a political viewpoint, the potential for activists to assert power and influence—social impact—is greatly enhanced.

Digital activism is frequently dismissed as passive, ephemeral, and superficial. Critics charge, for example, that social media may be good for exchanging information but not for strategic organizing. Further, critics note that social media may be good for generating awareness for social issues but not for deep and sustained political engagement. These same critics often overlook the substantive ways in which digital activism enables whole new repertoires of community building and political agency, not because of technology but rather because of the inventive ways activists use technology.

This new repertoire was never more evident than when young creatives came together online and offline to create what the *New York Times* called the first civil rights movement of the twenty-first century.

#FERGUSON

On August 9, 2014, a white police officer named Darren Wilson shot and killed an unarmed Black teen, eighteen-year-old Michael Brown, in Ferguson, Missouri. It was not the first time an unarmed African American was killed by the police that year. In fact, at least three hundred African Americans would be killed by police in 2014, the majority of them unarmed. But Brown's death triggered a local and national movement that signaled a turning point in American civic life.

Within a few days of Brown's death, Ferguson was dominating the news headlines, but the events there did not become a national story until they became #Ferguson, a social media–enhanced narrative and form of connected activism that was driven largely by young creatives. The protest in Ferguson placed a bright spotlight on the nascent Black Lives Matter movement and the vitality of connected activism.

#Ferguson was both a revelation and a revolution. What made #Ferguson a revelation was the rapid pace and intensity with which everyday citizens began reporting about the tragedy. From the beginning, the young creatives in Ferguson were not simply chronicling what had happened and what was happening between police and the largely Black working-class community, they were also developing a distinct point-of-view that was sensitive to the plight of Ferguson residents. In the words of African American millennials, #Ferguson was "unapologetically black." The reporting from the streets was designed to build support for local Black residents while also framing the militarized tactics of Ferguson police as excessive and oppressive.

What made #Ferguson a revolution was the degree to which young creatives became the primary source of news and information for the world. Social media–powered citizen journalism enabled activists to shape the larger political discourse around race, policing, and social justice in the US. More precisely, #Ferguson influenced how local, state, and federal law enforcement, media organizations, and elected officials, including the president of the United States, responded to the crisis. #Ferguson represented a power shift in our political and media culture and the political awakening of a generation.

The constant live updates via Twitter, Facebook posts, and video streams on Vine provided an intimate window into the epic struggle citizen-activists waged against law enforcement and elected officials for dignity and accountability. The most compelling photos and videos—almost all of them captured with smartphones—were shot from the point of view of Ferguson residents and activists. This was a radical departure from more traditional modes of news and information production, which are typically filtered through the lens of institutional elites such as law enforcement, elected officials, and professional journalists.

The first few days of social media documentation in Ferguson turned out to be critical. Ferguson residents built an early and compelling narrative that would spread and grow in terms of influence. By the end of the day of Brown's death, August 9, nearly 200,000 #Ferguson tweets circulated. By contrast, none of the major cable news channels—CNN, FOX, MSNBC— reported on the events in Ferguson that day. Local residents and activists broke the Ferguson story through their use of social media.

Two researchers from Northeastern University, Sarah J. Jackson and Brooke Foucault Welles, studied the first week of social media that flowed from West Florissant, the street where Brown was shot dead and which later became one of the flashpoints in the conflict between police and local residents. They examined over 500,000 tweets that were generated during the first week of the crisis that contained the keyword "Ferguson." Their findings are revealing.

The most influential Twitter users during that time were not journalists, elected officials, or leaders of civil rights organizations. Rather, they were Ferguson residents who took on the dual roles of citizen journalists and activists. Jackson and Welles call these citizen-activists "early initiators." They were among the first on the scene and the first to start posting information about the events that ensued after Brown's death. These early initiators were the most retweeted and mentioned among those who adopted Twitter to comment and report on what was happening in Ferguson.

This group of early initiators emerged as a grassroots elite, leaders and influencers in a network of activists and content creators that formed soon after Brown's death. The grassroots elite gained prominence as a result of how their storytelling resonated within a growing social network. Due to the mobile, visual, spreadable, scalable, and impactful nature of connected activism, the status of the members of this grassroots elite was elevated to a level that was comparable to more traditional elites, such as professional journalists and civil rights leaders. The young creatives in Ferguson provided a steady stream of information, on-the-ground reporting, and live updates that established an information infrastructure that grew in scope and influence.

Anyone paying attention to the #Ferguson political movement could see that the civic decline narrative did not apply to the young creatives mobilizing around Michael Brown's death. These were not millennials who were less engaged with news and political knowledge than previous generations of Americans. Rather, these connected activists embodied the civic innovation narrative, developing inventive, responsive relationships to news and political knowledge. #Ferguson established a new framework for how we think about the creation, circulation, and consumption of news and political knowledge in the age of social media.

One of the first tweets related to Brown's death was from ThreePharoah, a St. Louis–based rapper. At 12:03 p.m., roughly two minutes after Brown was gunned down, he tweeted, "I JUST SAW SOMEONE DIE OMFG."

One minute later ThreePharoah posted a photo of Brown lying facedown in the street with two white police officers standing over him. The tweet simply read: "Fuckfuck fuck."

ThreePharoah sent a few more tweets describing the scene as the crowd along West Florissant began to swell. Some of his followers tweeted back questions asking what had happened. Other followers used Twitter to ask him how he was holding up in the face of trauma. These exchanges typified the dynamic role that social media would play in Ferguson. From the very beginning, social media was, among other things, a real-time channel for reporting from Ferguson, spreading the perspectives of residents and activists, and building a community of material and emotional support.

In an interesting twist, the national news media followed the on-the-ground reporting from the army of young creatives in Ferguson who turned their mobile phones and social media into a source of citizen journalism, documentary television, and political activism. MSNBC's Chris Hayes told the *New York Times*, "This story was put on the map, driven, and followed on social media more so than any story I can remember since the Arab spring." This trend—citizens producing and circulating news via social media—was under way before Ferguson and reflects, more broadly, a collective turning away from the long-standing hegemony of legacy news media and their agenda-setting role.

This shift in reporting and civic storytelling had a profound effect on Ferguson, the nation, and, most significantly, the young creatives who were largely responsible for it all. Young activists would emerge from #Ferguson buoyed by the sense that they could influence the media and political elite.

After Ferguson, they recognized that connected activism could be a powerful lever for civic innovation, direct action, and social change.

#WOKE

To understand the vitality of connected activism, consider the political journey of DeRay Mckesson, the former Minneapolis school administrator who emerged as one of the celebrity activists from the tragedy in Ferguson. Mckesson followed the citizen protests in Ferguson through social media rather than the traditional news media. He was not alone. In our *Millennials, Social Media, and Politics* survey, 70 percent say that they get "a lot" or "some" of their news and information from social media. By contrast, 22 percent say that they get "a lot" or "some" of their news and information from print newspapers. The tweets, Facebook posts, and Vine videos that exploded from the streets of Ferguson resulted in an awakening of a sort for Mckesson. Black millennials frequently use the term "woke" to refer to a person who has developed a new awareness about something, usually a social issue, they were once oblivious to.

Mckesson would later assert, "We aren't born woke, something wakes us up, and for so many people, what woke them up was a tweet or a Facebook post, an Instagram post, a picture." The social media content generated by young creatives in Ferguson did more than wake Mckesson. The more personal accounts of what was happening in Ferguson gave him a connection to the movement that was extraordinarily personal and powerful.

Recalling the moment he decided to "go stand in solidarity with the people in Ferguson," Mckesson said, "It was 1 a.m. on August 16, 2014, and I'd seen the events unfolding in Ferguson via Twitter. And I waited until the morning and then called my best friend and asked him for his advice with regard to going down to St. Louis." Mckesson added, "I packed a small bag, put a status on Facebook saying that I was going and asking for somewhere to stay in St. Louis. Then I got in the car, drove nine hours, and ended up on West Florissant."

In Ferguson, Mckesson instinctively did what millennials do. He pulled out his smartphone and began capturing photos and videos of the protesters and their encounters with Ferguson police. He tweeted around the clock and posted hundreds of videos to Vine, giving followers a more intimate connection to what was happening during those grim days and nights in the streets. Within a few weeks Mckesson's social media activities—the nonstop tweeting and Vine videos—positioned him as a central node in the ecosystem of a growing political movement.

Mckesson and other activists transformed common social media practices—clever tweets, Facebook posts, memes, video clips, status updates—into dynamic forms of connected activism. In the midst of advocating for social justice, they pioneered new methods of political communication and expression. Much of it was improvisational, as Mckesson explained. "There is no one way to do this work. There's no one way to be someone who cares about justice or equity," he said. "There's no one way to use tech platforms. If we had used Twitter the way that all the articles say that you use Twitter, we wouldn't be here. We use it in a different way. . . . You think about the beginning of the protests. It was before . . . you could upload videos on Twitter. We were really patchworking platforms to make them work for us."

As the movement in Ferguson evolved, social media became a resource for connecting to other activists, coordinating demonstrations, and sharing information. "Twitter," Mckesson would later say, "was how I processed [my experience in Ferguson]. I quickly understood Twitter to be a really powerful organizing tool, and we used it to bring people together, to challenge narratives that were untrue, to push people to think differently. It became a real force."

For many young creatives, social media was an opportunity to keep people connected not only to one another but also to the events that were unfolding in the streets of Ferguson. Twitter was a resource for reporting and witnessing. "I remember when Trayvon Martin died, there was no news, and I just didn't know what was true or not. I didn't want that to be the story of Mike Brown," Mckesson told the *Advocate*. In direct contrast to the civic decline narrative, social media was smartly leveraged by young creatives in Ferguson to amplify voices of political dissent. Many of the social media posts captured images of military-style policing—the use of tear gas, police dogs, and armored tanks—that were shared with people all across the world.

The use of social media in Ferguson had a strong social component too. For Mckesson, Twitter was a tool for community building as well as political activity. In interviews, he liked to say that "Twitter was the friend that was always awake." No matter the time of day he posted to Twitter, someone was always online and likely to respond. In direct contrast to the civic decline narrative, young creatives used social media during the events in Ferguson to invest in social capital. Each time Mckesson reached out on Twitter and found someone awake, he strengthened his ties to the woke community he was a part of. Mckesson realized the vast potential of social media to build his social capital. He used social media to reach out to people, cultivate social connections, and find the material and emotional resources he needed during challenging situations. The community and connections he built online also offered support in the face of internet trolls, bullies, and hate speech.

Mckesson's political journey compelled him to think about the perils and possibilities of digital activism. He was annoyed by the charges that "internet activism" was passive, ephemeral, and shallow. Critics often dismiss internet activism as clicktivism, a term typically used to deride digital activism in general and millennial activism specifically. As he reflected on his own personal experience, Mckesson noted, "I never criticize people who [others] deem to be Twitter activists, or hashtag activists, because I know that telling the truth is often a tough act, no matter where you tell that truth. I think that's important." Mckesson added, "I think that we'll continue to see the platforms push and redefine the way we organize."

Mckesson is among a growing network of young creatives who are shattering derogatory notions of digital activism by pioneering new models of political engagement. Digital activism is not a monolithic enterprise or a substitute for deep engagement in civic life. Rather than retreat from civic life, young activists like Mckesson are expanding the terrain of engaged citizenship and the repertoire of practices we associate with civic life. In our survey with NORC at the University of Chicago we noticed an interesting relationship between social media use and civic engagement. Millennials who were most likely to post political or social issues content via social media were also more likely to be engaged in civic-related activities offline such as voting or attending a political rally. For them, social media was not a passive form of political participation but was instead associated with more active forms online and offline. These activists are not substituting connected activism for real-world engagement; they are using the power of connected activism to complement real-world engagement. Young activists like Mckesson certainly embody this finding in our data.

While it was Mckesson's inventive use of social media that catapulted him into political celebrity, he did more than tweet about injustice. Mckesson essentially lived in Ferguson for several months, joining protesters to confront law enforcement and city officials. He then joined activists in cities like Baltimore and Baton Rouge to protest the killing of unarmed Black citizens by police. David Axelrod, longtime political insider and adviser to

President Obama, invited Mckesson to lead a seminar on social media and activism at the University of Chicago's Institute of Politics. Mckesson, a Baltimore native, even ran for mayor in his hometown. Beginning in Ferguson, Mckesson collaborated with a diverse team of activists, artists, designers, and data scientists to build a new kind of political movement that reflects many of the signature features of the new innovation economy. Young creatives like Mckesson wield a form of political activism that is networked, tech savvy, and deeply committed to social justice.

SOMETHING BIGGER

#Ferguson marked the beginning of an explosion of street protest and savvy social media engagement that emerged in the days and months following the death of Mike Brown. In retrospect, #Ferguson was a precursor to something bigger. Some of the young creatives involved in #Ferguson began to think about their social justice work the way aspiring entrepreneurs think about a start-up, and that meant mobilizing their resources to build an enterprise that they could grow, iterate, and deploy to disrupt, in their case, the political status quo. As their model of political activism evolved, it began to embody some of the core features of the new innovation economy. This was certainly true with Samuel Sinyangwe, a Stanford University graduate and policy analyst.

Like so many other millennials, Sinyangwe was struck by the rise of the Ferguson protest and how it "woke" a generation. When he saw Mckesson's reply to Oprah Winfrey's criticism that Black Lives Matter activists did not have specific asks or goals, Sinyangwe reached out to Mckesson via Twitter. "I replied to the tweet saying that I could help develop a policy agenda that implements these demands in practice. I didn't know who De-Ray or anyone was," Sinyangwe recalled. "As a policy analyst, I wanted to contribute policy."

Sinyangwe believed that if the advocacy work of Black Lives Matter was going to spark real changes in the way law enforcement polices Black communities, the activists needed to persuade lawmakers to act. The best way to do that, he thought, was to tell a data-driven story. Sinyangwe identifies as a data scientist, but he is also a data activist. He represents an emerging group of talented professionals who are beginning to think inventively about the ways data can be mobilized to inform and support movements that challenge systems of social and economic inequality. In the era of "big data," we are learning that the collection of massive amounts of information can and has been deployed in ways that reproduce disparities in health, education, up-ward mobility, and criminal justice. The way algorithms are developed, the analysis of large data sets, and the application of artificial intelligence reflect the degree to which the management of data serves as a source of power, new capitalism, and social control.

Data activists believe in the power of data to tell stories that have the capacity to induce political action and policy change. Sinyangwe believed the next evolution in the movement to secure Black lives was the use of data as a tool for mobilizing more strategic and persuasive forms of political en-gagement. "People in positions of power and influence are more receptive to data than stories. In their positions they hear all kinds of stories from all kinds of people, and they have to sift through what the trends are in order to set policy," Sinyangwe said.

After a series of conversations with Mckesson and St. Louis–based activ-ist Brittany Packnett, Sinyangwe decided to conduct an analysis of police killings in the US. If he and his fellow activists were going to advocate for policy shifts and greater accountability among police, it was important to have a precise understanding of the scope of the problem. For example, how many people do the police kill a year, and what percentage of those people is African American?

When Sinyangwe went to collect the data, he was surprised by what he found—nothing. More precisely, he discovered that none of the federal agencies that we might expect, including the Department of Justice, the FBI, and the Centers for Disease Control, maintain records of police use of force. This was true even though one of the requirements laid out by the Violent Crime Control and Law Enforcement Act of 1994 was that the attorney general's office is supposed to publish an annual summary of police activities, including the excessive use of force. But as he searched, Sinyangwe realized that there were no standardized protocols for collecting and systematically organizing data about the activities of the police.

Eventually, Sinyangwe turned to three citizen-driven crowdsourced da-tabases to get the most comprehensive information on the scope of police use of force. KilledbyPolice.net, the US Police Shootings Database, and FatalEncounters.org highlight the persistent efforts of everyday citizens to keep the public informed about the use of deadly force by police. These civic-oriented enterprises were doing what the federal government was not doing—maintaining an annual accounting of police killings across the coun-try. It did not take Sinyangwe and his collaborators long to realize that the databases were useful. Combined, they represented roughly 90–95 percent of the total police killings in the US. Few people knew that these sources of data existed. Sinyangwe told a group gathered for a 2015 conference on Data & Civil Rights that the databases "have been here all along. It's just no one had taken the data bases, merged [them], filled in the gaps, and made sense of it to the world." That is precisely what he and his collaborators decided to do. Sinyangwe explained, "We could tell the story in another unique way, a way that can be especially appealing to policy makers."

After collecting and analyzing the data, Sinyangwe and his colleagues had to decide on the best way to share their findings. What story would they tell, and, equally important, how would they tell it? After careful deliberations, they decided to build a web-based project called Mapping Police Violence. Working with a team of artists, designers, and web developers, the team fed the information from their unique data set into Carto, a location data intelligence software solution that, among other things, visualizes geo-tagged data. They designed graphics and charts to illuminate some of the more striking findings from their analysis. The site that they built went live in March 2015—seven months after Brown was killed—and was immediately recognized as a technical and civic achievement. News organizations like the *Washington Post* and the *Los Angeles Times*, as well as tech news organizations like *Fast Company*, the *Huffington Post*, and *TechCrunch* featured the project in their reporting.

Mapping Police Violence was also eye-opening for the team that created it. "One look at that map, in two seconds, you knew this was happening everywhere. It wasn't just a Baltimore problem or a Ferguson problem," Sinyangwe told *Fast Company*. The visualization data compiled by the team supported a rallying cry that was echoed by many activists regarding the use of deadly police force in the US: "Ferguson is everywhere."

The empirical data was equally compelling. Take, for example, their analysis of 2015 data on police killing from the sixty largest police departments in the US. In 2015 fifty-nine of the sixty police departments killed civilians. The rate of police killings exceeded the national homicide rate in several cities, including Bakersfield, Oklahoma City, Oakland, New Orleans, Indianapolis, and St. Louis. Of the sixty departments reviewed, only one, the Riverside (California) Police Department, did not kill anyone in 2015. The data revealed that Blacks were disproportionately more likely to be killed by police than any other racial or ethnic group. Among the people killed in the top sixty police departments in the nation, Blacks made up 41 percent of the victims, even though they were only 20 percent of the total population in these specific jurisdictions. Forty-one of the sixty police departments disproportionately killed Black people relative to the population of Black people in their jurisdiction. Alarmingly, fourteen departments killed Black people exclusively, including St. Louis, Atlanta, Kansas City, Cleveland, Baltimore, Boston, and Washington, DC.

For Sinyangwe, Mapping Police Violence was precisely the kind of data-driven story that he envisioned telling to support the movement to secure Black lives. The findings revealed in Mapping Police Violence led Sinyangwe, Mckesson, and Packnett to launch Campaign Zero, a civic initiative that included ten specific policies to reduce police killings in the US, such as the use of body cameras for police and a more robust training regimen for police officers. Campaign Zero was the team's first explicit move into policy-oriented activism. A mix of stakeholders, including police departments, elected officials, and high-profile political candidates consulted the policy ideas crafted by Campaign Zero, marking key shifts in the identity and influence of a movement that began in Ferguson.

A civic enterprise like Mapping Police Violence required an approach to political activism that was collaborative, tech savvy, and empowered by a diverse network of talent. It also required activists to become more innovative and even entrepreneurial in their desire to be agents of social change.

BUILDING A CIVIC START-UP

Shortly after releasing Mapping Police Violence, Sinyangwe, Mckesson, and Packnett formed a new political organization, WeTheProtesters. Their post-Ferguson efforts had increasingly focused on translating their experiences, knowledge, and political capital as activists into a more sustainable civic enterprise. Efforts like these were designed to catalyze the momentum generated by Black Lives Matter.

The young activists like Sinyangwe, Mckesson, and Packnett that emerged from Black Lives Matter had ambition, but that was not enough to create a political movement built for the twenty-first century. The evolution of their ideas about political engagement prompted them to begin thinking as much like a tech start-up as a civil rights organization. Naturally, this meant thinking about how their technology and social ingenuity could empower their civic aspirations. In an effort to launch a higher-capacity organization, they took a path that a growing number of aspiring start-ups take—they applied for admission to an accelerator.

Since 2008, the accelerator model has been growing at a steady clip in the US. Accelerators are designed to ramp up a start-up's learning, networking, execution, and go-to-market preparation in a fixed amount of time, usually about six weeks. I visited a number of accelerators during my fieldwork and was struck by several features. In many ways, accelerators are like boot camp for entrepreneurs. Their primary aim is to help start-ups scale their idea for a product or service. Some of the most recognized companies in the digital economy—Airbnb and Dropbox—benefited from the accelerator model.

The typical accelerator provides a physical space for start-ups to build an idea into something tangible. Accelerators are also a source of social and financial capital. For example, accelerators often connect aspiring start-ups to a vibrant network of entrepreneurs who can offer, among other things, business, strategic, and technical expertise. Additionally, accelerators introduce promising entrepreneurs to a network of angel investors and venture capitalists, the financial backbone of the innovation economy.

WeTheProtesters was accepted to a San Francisco–based accelerator called Fast Forward. Whereas most accelerators are designed to help launch for-profit companies, Fast Forward was one of the first to specialize in cultivating nonprofit enterprises. For traditional accelerators, a key measure of a company's success is the value of financing rounds or lucrative exits. Accelerators that build nonprofit organizations establish different metrics for success. For example, Fast Forward measures success in terms of the number of lives their companies have impacted, the lines of code written, and the percentage of founders who are women and from diverse racial and ethnic groups. During their time at Fast Forward, the founders of WeTheProtesters had the opportunity to sharpen their strategic vision, grow their social network, meet and learn from other social impact enterprises, and prototype their idea for a digital civil rights movement.

Unlike in the for-profit sector, in the nonprofit sector there is no venture capital infrastructure. The leaders of WeTheProtesters knew that the timing for their idea—a millennial-driven political movement—was right, but they would need to find more creative ways to prepare to go to market and to scale their enterprise for the immense audience of activists they wanted to reach. How could they build an operation that effectively mobilized the widespread desire among many millennials to get involved? As they thought about their political future, they discerned that the scope of the work was far greater than any one organization could manage. The number of people either directly involved with or at least sensitive to the claims and concerns of Black Lives Matter was substantial. The challenge that WeTheProtesters faced was designing a mechanism and a call to action that could effectively mobilize this untapped resource into a base of direct action, influence, and political power. "There are not enough full-time activists and advocacy organizations to handle the immense demand of the moment," Sinyangwe said, "so we have to figure out how to build the capacity to the millions of people who want to get involved to be as good as the professionals."

A TIME TO ACT

WeTheProtesters' desire to build what some called the first civil rights movement of the twenty-first century required ingenuity, verve, and the ability to tap talent that existed beyond their small circle of leaders. The contributions of Aditi Juneja, a former New York University law school student, are a brilliant example of the possibilities of connected activism in a networked society.

By her own admission, Juneja was late to the Black Lives Matter movement. She knew the group existed, but she struggled to articulate anything specific about its practical goals. In her third year of law school, Juneja reached out to WeTheProtesters via social media, indicating that she had a background in government and an interest in policy. She was struck by the fact that a group of young activists were not only making noise; they were also making a difference. In 2016, soon after corresponding with Sinyangwe, Juneja began serving as an informal adviser to WeTheProtesters in between her studies at NYU Law. Americans would elect a new president that year.

Juneja paid close attention to the presidential campaign. She followed the vastly different policy positions of the two major candidates, Hillary Clinton and Donald Trump. Like many Americans, she was surprised by the election of Trump, calling it a wake-up call for her and her colleagues at WeTheProtesters. "After the election, we knew we needed to shift focus," Juneja told me. Trump's policy pronouncements, on everything from immigration to health care, were a frontal assault on WeTheProtesters' values of equity and social justice.

A few days after the presidential election, Juneja took action. "I started making a Google Doc and tables to keep track of different policies. It started out as an attempt to educate myself." She was interested in how the looming shifts in the policy landscape would impact those who were most vulnerable, socially and economically. "There is a lot of chatter about politics in the news media," she explained to me, "but not much attention to how politics and policy actually impact people's lives."

As she began studying the nuances of the political and legislative process, Juneja became extraordinarily knowledgeable about policy. During our conversation, she spoke about things as varied as budget reconciliation, cloture, and the Affordable Care Act. Juneja believed strongly that it was important to build a platform that could help educate aspiring activists. "I was a law student and someone who paid at least a little attention to the connection between politics and policy," Juneja noted, adding, "and if I did not understand what was happening in the policy space, I was convinced that many others did not either."

After populating the Google spreadsheet with policy-related content, Juneja shared it with Sinyangwe. "Sam liked it, and we both thought, 'This should be made public.'" Juneja and some of the WeTheProtesters team began brainstorming the best way to leverage her policy document into a form of political action. She explained to me that while the project was motivated by the 2016 presidential election, it was not anti-Trump. Rather, she wanted to build a platform that was pro-equity and pro-social justice. Juneja's policy document became the material for a new open-source, wiki-style platform that embodied WeTheProtesters' vision to build capacity for a twenty-first-century political movement that was digital, networked, creative, and, most importantly, capable of inspiring direct political action. They called the platform the Resistance Manual.

PROTOTYPING THE CIVIC FUTURE

If there was one social platform that was synonymous with Black Lives Matter, it was Twitter. Despite the 140-character constraints at the time, Twitter was a multipurpose channel. Activists used it as a platform for organizing, broadcasting, and collective mobilizing. Twitter inadvertently became the infrastructure for the movement. But as WeTheProtesters began to think about building a more sustainable movement and an organizational structure, Twitter had some notable limitations. "What it [Twitter] hasn't done effectively is preserve the knowledge base outside of the immediacy of the moment," Sinyangwe explained in an interview with the *International Business Times*.

The Resistance Manual would be a different kind of social platform, he said: "[It] promotes crowd sourcing mentality and collaboration at scale, but does it in a way that saves the content and allows you to dive deeper and really become informed." If Twitter was about staying informed in real time, the Resistance Manual was about staying informed over time. The inspiration for the design of the Resistance Manual came from the most collaborative information resource in human history: Wikipedia. The Resistance Manual runs on the same free software as Wikipedia, MediaWiki. From a design perspective, the Resistance Manual looks and feels like Wikipedia. The layout and organization of content follows the wiki model, making it both familiar and user-friendly. Equally important was the concerted effort to capture the collaborative spirit of Wikipedia. "When you think about Wikipedia, they're designed to crowdsource contributions from people effectively and to build a knowledge base that's greater than any one group or organization," Sinyangwe said. "It's reflective of the body of knowledge that's out there."

The wiki model was responsive to the organizational and financial constraints that WeTheProtesters faced. The fledgling civil rights organization was ambitious, but it lacked the two things that are critical to mobilizing a political organization: money and staff. The wiki model, like so many of WeTheProtesters' other endeavors, became a clever workaround in the face of limited resources. What it may have lacked in financial or human capital, WeTheProtesters made up with the accumulation of social and reputational capital. Whereas social connections lead to social capital, the respect and admiration inspired by activism result in reputational capital.

When it came time to execute the wiki, WeTheProtesters mobilized their deep social connections to recruit talent. Social media helped to widen their social network. Research has shown that it is not necessarily the size of a social network that matters, but rather the diversity of the people in that network. Social network scholars refer to this as "network extensity." Launching a digital platform required a deep reservoir of talent—designers, coders, artists, researchers, writers, and policy analysts. By growing their social capital, largely through social media, WeTheProtesters was able to access the human capital it otherwise lacked.

WeTheProtesters had swag too. The organization's brand was recognizable and reputable among young, established, and aspiring activists. In the wake of Ferguson and other high-profile police killings, Mckesson had become a political celebrity. He made appearances on *The Late Show with Steven Colbert* and *The Daily Show with Trevor Noah*. Johnetta Elzie and Packnett had been featured in a number of news media outlets. Elzie, a St. Louis native, used her razor-sharp intellect to bring attention to the racial injustice she had witnessed in Ferguson. She was known as a "Day 1," a reference to those who began protesting in Ferguson the day Brown was killed. Their smart use of social media—everything from Black Twitter to meme culture—reflected a generational ethos that was transforming the cultural and communication landscape. The social media presence of the WeTheProtesters planning team, simply put, enhanced their reputational capital and appeal to young people.

Juneja's vision for the Resistance Manual was straightforward: policy explanation and connections to activism for as many people as possible. Her "napkin pitch" went something like this: "Get educated. Get organized. Take Action." The pitch summarized the three things the Resistance Manual was designed to do: First, educate about specific policy areas. If activists were going to spark change, it was critical that they be knowledgeable about the political process and specific policy areas. Second, the Manual was designed to function as a hub that curated the knowledge, insights, and tactics activists could use. Finally, the Manual was designed to offer local activists concrete pathways into direct political action.

For its launch, the Resistance Manual offered a mix of materials, including crisis resources, tools of resistance, and essential readings. For example, the executive orders on immigration issued by Trump shortly after he took office thrust many immigration activists into crisis mode. Undocumented adults needed to know where they could go to get legal assistance and support for their children if they were deported. The Resistance Manual offered a number of specific tactics activists could include in their civic toolkit. Some of the tactics included strategies for working with the media, including how to write op-eds or give an interview. The essential readings list compiled books, articles, and syllabi that allowed people to access a wide range of ideas and philosophies related to the history of protest, social movements, and inequality.

MAKING SHIT UP OR FIGURING THINGS OUT?

At its core, the Resistance Manual resembles some of the creative and entrepreneurial endeavors that I observed throughout my research. There is no dedicated office space. The bulk of the work—research, data collection, strategizing, policy discussions—happen primarily through online channels. Juneja was based in New York, but the community of volunteers that she worked most closely with lived across the US. "We had people from Oregon, New York, Missouri, North Carolina, Texas, really all over the world," Juneja recalled.

Like many other bootstrapping innovators, WeTheProtesters turned to online tools to help make their work more efficient, productive, and manageable. Juneja told me that they relied heavily on Slack, the cloud-based business application software, in those first few months of the Manual. The team of volunteers for the Resistance Manual used Slack to coordinate, communicate, collaborate, and ideate. "It was great for getting things done," Juneja told me. She was overwhelmed by the enormous amount of work the Manual required. "I was working nonstop, trying to manage the site, overseeing new content creation, and responding to more and more inquiries." Once she set up a leadership team and developed Slack subgroups, the actual job of organizing the influx of correspondence, questions, and volunteers became more doable.

With Slack they could create subchannels and subconversations that allowed for a more efficient exchange of information, collaboration, and problem-solving. For example, the members of the leadership team who focused on immigration organized through a specific channel that allowed them to share relevant information and devise a strategy for curating that information. In the early days of the Manual, immigration was a hot topic, largely because of the swift action Trump took shortly after becoming president. People needed information right away that addressed questions such as "What are my rights?" and "Who can I consult for legal advice?" Immigration activists from communities across the US needed help identifying specific tactics they could use to support immigrant families and communities in crisis.

Within just a few days of the launch, the Manual generated a few hundred thousand unique users and several thousand edits. According to Juneja, the users included individual citizens, nonprofit organizations, ex-patriots, and activists from other countries. There was, for example, the mom in Chicago who used the Manual to inform her kids and spark conversations with them about racial, political, and policy matters. The conversations initiated by this mother inspired one of her kids to start a WeTheProtesters club at her high school, a likely pathway into civic engagement for this group of young citizens.

The Manual also became a widely respected tool in the larger "resistance community" that grew rapidly after the election of Trump. As more social impact organizations form, competition between them is inevitable. Among other things, groups vie for funding, visibility, talent, and, of course, influence. Juneja believes that rather than competing in the "political resistance nonprofit space," the Resistance Manual emerged as a neutral tool. She noted that when she attended conferences and convenings with members from the resistance community, they consistently commented on the value of the Resistance Manual. "They view it as a resource for strengthening their knowledge in key policy areas," she explained. Soon after its launch, Juneja realized that the Resistance Manual had the potential to be a pivotal node in a political ecosystem where youth activism and the policy landscape might be transformed.

The Manual's audience was global too. "People and organizations from India, South Africa, and the UK reached out to us," Juneja said. As a result of #Ferguson, the activists at WeTheProtesters became global icons of digital age political activism. Aspiring activists from other countries wanted to learn more about the strategy, organization, and ideas that invigorated the Resistance Manual. Juneja's conversations with global activists were substantive. When she spoke with activists from India, Juneja pointed out that the lower levels of internet penetration in the country might limit the value of a digital toolkit like the Resistance Manual. She noted that pamphlets and radio addresses might be more useful in communities facing digital access and literacy issues.

Juneja's side hustle—building the Resistance Manual—eventually became a full-time gig. Roughly two hundred people—lawyers, teachers, designers, activists, and students—signed up as volunteers soon after the Manual's launch. Somebody needed to quickly devise and manage a system that could utilize this volunteer talent effectively. Additionally, someone needed to create and supervise a process that could manage the crowdsourced content that populated the platform. The Resistance Manual was Juneja's idea, but it was never her plan to be the lead organizer of the platform. "I never expected to be that person," Juneja told me.

While attending NYU law school—from which she later graduated—Juneja worked as many as seventy hours a week on the Resistance Manual. "I did not go to many classes. Really, I did not do much else." As Juneja thought about the massive effort that was required to build early momentum for the Manual, she reflected, "This movement called for something different." Before long she was managing the flock of volunteers, writing and curating content for the site, conducting research, and networking with other resistance organizations. In her words, "I was just making shit up as

I went along." There was no manual for how to organize a digital political movement. She read a few books and studied a handful of organizations, but this was new terrain for her and for WeTheProtesters.

What Juneja described as "making shit up" could also be characterized as "figuring things out." It turns out that the ability to adjust on the fly, grapple with complexity, manage uncertainty, and figure things out are among the most crucial skills in an economy powered by innovation. The ability to see the need for a product, a service, or an idea and figure out how to deliver it is a skill. The team at WeTheProtesters saw a need—catalyzing a burgeoning social movement—and proposed a solution for tapping and unleashing its political potential: a crowdsourcing platform that built community, knowledge, and tools for direct action. One of their innovations was the scale at which they attracted a diverse collection of talent to build a political organization that was much larger and potentially far more impactful than any traditional organization they could have created.

THE CHALLENGES OF CONNECTED ACTIVISM

If there was one thing that Juneja wanted me to understand about the Resistance Manual, it was how hard it was to bring the idea to fruition. "All of the work that we did sounds romantic, but it was not. It was really, really hard," she said. There were times when she questioned if they could realize the vision. "It's not pretty; you make mistakes. I yelled at people, neglected family and friends," she acknowledged during our conversation.

I was struck by how much Juneja learned about civic innovation largely through the hands-on experience of making the site and then going public with it. She led the effort even though she had no experience building a civic media platform. Like many of the enterprises discussed in this book, the Resistance Manual was a product of tech ingenuity, social capital, responsiveness to change, and entrepreneurial hustle.

Juneja and the leaders from WeTheProtesters realized that activists around the country needed a network and tools that could empower their desires for activism, sometimes on a local scale, sometimes on a national scale. As a result of the spreadable and scalable features of connected activism, the Resistance Manual influenced activists from other countries too. "People around the world reached out to us," Juneja proudly recalled. "They wanted to learn from us in order to do something similar in their country." When Juneja met with some of these activists and shared her insights about using digital technology to organize political movements, her actions illustrated the sharing economy that is such an integral part of the innovation economy young creatives are building. Rather than charge global activists a fee to acquire insights from WeTheProtesters or to access the Manual, Juneja wanted the resource she helped to create to be accessible, spreadable, educational, and impactful. While it may have cost WeTheProtesters financially, the decision to make access to their wiki-based platform free invited the expertise, ideas, and community-building that brought their vision of a more dynamic, effective, and inclusive civic future to life.

DETROIT HUSTLES HARDER

Why the Motor City Matters in the New Innovation Economy

A fter decades of steady and striking decline, the white population in Detroit is growing. Demographers started noticing an uptick in white residents in 2010. In 2014 the uptick reached what the population experts call
"statistical significance." According to William Frey, a noted demographer
with the Brookings Institution, young singles and married couples without
kids are driving the white population growth. This is after decades of turbulent history—often related to race—that followed the decline of Detroit's
famed and once dominant automobile industry.

Today there is a sense of optimism about Detroit because it offers something that very few major cities offer young creatives—an opportunity to
build their own future. The one certainty that millennials face in today's
society is economic uncertainty. The prospects of homeownership, full employment, and economic security are not very good. It is a combination of
affordable housing and a growing tech and entrepreneurial economy that
powers the influx of young whites in Detroit. I call these transplants the
"young adventurers," as they are eager to kickstart their lives in a city where
opportunities are more than a mirage. One young adventurer explained his
decision to launch his start-up in Detroit this way: "You can make a bigger impact quicker. The rent is cheaper. We get media coverage simply because we're a company that's actually choosing to be in Detroit." The young
adventurers are part of a vibrant ecosystem that is forming in the center
of downtown, attracting a new generation of designers, techies, and entrepreneurs who are driving what one magazine calls the "Brooklynization of

Detroit," a reference to a once-forgotten city suddenly being adopted and transformed into a magnet for young creatives on the come up.

The "come build your future" narrative is more than a slogan. It is a systematic strategy that has been embraced and employed by the city's financial and political elite as part of the pivot toward a new and more forward-leaning downtown economy. For longtime Black Detroit residents, the "come build your future" narrative is a source of genuine tension. It suggests that their efforts in the city—indeed the very grit they have displayed by staying in Detroit when so many left—are largely undervalued and undercapitalized compared to the Downtown district. There has been no comparable investment of public-private partnerships to revitalize the parts of the city that are overwhelmingly Black, poor, and isolated from high-opportunity areas. The "come build your future" narrative suggests that Detroit will be saved by transplants who are mostly white, upwardly mobile, and highly educated. It is as if longtime residents are mere casualties of the city's buoyant future.

MOTOR CITY MAKEOVER

The three core forms of capital—financial, human, and social—that drive all innovation hubs power Detroit's entrepreneurial ecosystem too. The ability to supply potential high-growth companies with the financial capital they need to build their teams, products, and capabilities is critical to building an innovation hub. The state of Michigan has seen a recent upswing in venture capital funding, and Detroit has been a direct beneficiary. Growing venture capital funding in downtown Detroit is being channeled into sectors such as information technology, life sciences, and mobility. Similar to venture funding in other locales, Detroit's culture of financing innovation tends to favor entrepreneurs who fit a certain profile—usually white, male, college-educated, and presumed to be tech savvy.

The tech and business elite in downtown Detroit has been especially active in recruiting and cultivating human capital. In its *2017 Scoring Tech Talent in North America* report, CBRE Group, Inc., a global investment firm, looks at a number of unique metrics to rank the top fifty cities in the US and Canada according to their ability to attract and grow tech talent. CBRE ranked Detroit number twenty-one among cities attracting tech talent, placing them ahead of much bigger cities such as Philadelphia, Los Angeles, and Houston. Between 2012 and 2016, five cities grew their tech labor pool by more than 40 percent. Four of the cities—Raleigh-Durham, San Francisco Bay Area, Atlanta, and Portland—came as no surprise. The fifth, Detroit, is an unlikely member of that list. Once not even an afterthought among young creatives, Detroit is now, according to CBRE, a "brain gain," a city in which the number of people hired in the tech labor market exceeds the number of tech degrees awarded in the market.

Downtown Detroit is also rich in social capital, an asset that provides a vibrant network of social ties, support, and the sharing of ideas, information, and expertise. Downtown Detroit is home to a surging number of coworking spaces, start-up incubators, accelerators, and networking events that foster a dynamic web of connections and interactions that in turn cultivate new relationships, collaborations, and economic ventures. When Dan Gilbert, the billionaire founder of Quicken Loans, began investing in downtown Detroit, this is certainly what he had in mind—the creation of a lively hub of companies, talent, knowledge spillover, and entrepreneurial energy serving as the social engine for innovation and economic growth.

Detroit, however, continues to be the poorest big city in the US, with 39 percent of the population living below the nation's poverty line. The majority of the city's poverty is concentrated in Black neighborhoods. The arrival of young adventurers continues to exacerbate the racial and economic inequalities that are a striking feature of the "new Detroit." The distribution of positive demographic and economic indicators, such as growth in population, job creation, and per capita income, has largely been restricted to the 7.2 square miles that comprise the Downtown and Midtown districts. These trends, as one study notes, are in "stark contrast to what is happening in the other 95% of Detroit."

Black Detroit residents see a downtown area accelerating toward economic vitality; meanwhile, Detroit schools are among the most distressed in the nation, ensuring that homegrown talent will never be included in the new Detroit. A 2014 economic report prepared by the Kirwan Institute for the Study of Race and Ethnicity, a research and policy think tank at Ohio State University, corroborates the concerns expressed by many Black Detroit residents. The researchers at the Kirwan Institute use an "opportunity mapping analysis" to assess the quality of life in metropolitan areas across the US. Thriving schools, robust employment, higher-income neighborhoods, and low crime rates characterize high-opportunity areas, on the one hand. Declining schools, anemic employment, lower-income neighborhoods, and high crime rates characterize low-opportunity areas, on the other. The researchers found that more than half, 53 percent, of African Americans in Detroit live in low-opportunity areas of the city. By contrast, 36 percent of whites live in low-opportunity areas. The new Detroit is working for some, but not most. Rather than diminish the sharp racial and economic disparities that have long defined the city, the new Detroit is reproducing them.

A DIFFERENT KIND OF INNOVATION ECOSYSTEM

Even as Detroit's downtown economy has become a source of increased media attention, another kind of innovation ecosystem has been forming in its shadows. The model is one the city should seek to replicate and scale locally. It began when a thirty-thousand-square-foot building in Corktown, a historic neighborhood located near downtown Detroit, caught the attention of Phillip Cooley, a local restaurant owner. Like so many other abandoned structures in the city, the building had gone up in 1935, during the city's glorious industrial heyday, when it had been a printing facility that supplied materials for the auto industry. But now it was going through foreclosure. Cooley purchased it from the bank for $100,000—a bargain if there ever was one.

The immediate challenge was figuring out what to do with the massive structure. While brainstorming ideas of what they could do with the space, Cooley and his partners kept coming back to one fundamental idea: What if the building could become an anchor institution for a new kind of innovation ecosystem—one that benefited a wider community of creative types rather than a select group from the tech, educated, and entrepreneurial elite? They decided to offer inexpensive work space to socially conscious designers, entrepreneurs, and artists.

When it was time to settle on a name for the new initiative, they chose Ponyride. Karen Bordine, who served as Ponyride's initial executive director, explains the inspiration for the name. "We want to bring people back to the time when they were kids—curious, unbridled and more creative. Everyone wants a pony ride. You might fall off, but you just brush off your knees and jump back on." Social entrepreneurship, Team Ponyride discerned, requires having the nerve to get up after being knocked down and trying again in the face of adversity.

Ponyride's goal from the start was to follow a different model than that pursued by the architects of Detroit's downtown innovation hub. In order to achieve a unique goal like this, the design principles that guided the construction and execution of the model had to be unique. Two of Ponyride's core design principles—diversity and education—were baked into the model from the very beginning rather than inserted later as an afterthought. This, of course, stood in sharp contrast to more conventional innovation hubs like Silicon Valley or the one emerging in downtown Detroit.

An aspiration among the architects of Ponyride was to build a space that thrived, in part, on the presence of a broad spectrum of talent, ideas, and approaches to innovation. This meant that the creative types and entrepreneurs using the space would look different from and think very differently than their counterparts Downtown. According to Ponyride, 60 percent of its entrepreneurs are women and/or persons of color. Another key feature in Ponyride's vision to build a more diverse innovation economy is reflected in its desire to grow the social capital of local entrepreneurs. This includes cultivating an ecosystem in which members of the community at large are empowered to participate in the innovation culture that Ponyride labors to create.

In a 2016 report outlining Ponyride's vision, the founders explained that it was not enough to have tenants who hired Black employees. They recognized the need to create opportunities for entrepreneurs and business owners of color, who could then empower others like them. This strategy, fostering an innovation ecosystem that includes and benefits the wider community, is an attempt to grow the social capital of local Detroiters. The idea is based, in part, on the theory that the best way to build a diverse and inclusive innovation economy is to recruit and support diverse talent who can serve as role models, information channels, and points of entry into the innovation economy to encourage the flow of diverse talent. By cultivating Black entrepreneurs, Ponyride hoped to inspire the development of additional Black entrepreneurs.

Ponyride's management may have seen their model as experimental, but the effectiveness of strategies like theirs is supported by social science. In this case, they were building, likely unknowingly, on research that suggests that the behavior of those around us can be contagious. Social networks have the capacity not only to shape behavior—what we do—but also to shape norms—what we are expected to do.

The Ponyride model also establishes an education requirement for each of its entrepreneurs. Erin Patten, a Ponyride staff member, told me that Ponyride's entrepreneurs "are required to share their knowledge with the local community." Each tenant must provide six hours of classes or other educational services each month. This design principle fulfills part of Ponyride's mission: to practice community engagement. This education and outreach come in many forms. For example, some tenants might share tips with community members in a class related to launching an entrepreneurial idea. In other instances, a tenant might teach a specific trade or skill related to their domain of expertise in crafts, filmmaking, or furniture making. The targets of the outreach could be adults looking for a career pivot or youth looking for role models.

The idea to be connected to and engaged with Detroit neighborhoods was incorporated into the Ponyride model early on, thereby establishing a purpose and an ethos that was markedly different from the model of innovation under construction in downtown Detroit.

TURNING THE FORECLOSURE CRISIS INTO A COMMUNITY ASSET

The foreclosure crisis has been an opportunity for Detroit's business and tech elite. The crisis made many of the city's buildings available at cut-rate prices. Dan Gilbert, of Quicken Loans, was a main beneficiary of these deals. Since moving Quicken Loans headquarters to downtown Detroit in 2010, Gilbert has acquired more than ninety properties in the downtown area. But there were others who also turned the crisis into an opportunity. The growth of coworking spaces, art and design studios, coffee shops, and trendy restaurants illustrates how the foreclosure crisis has led to a Downtown innovation economy built largely by and for a business, entrepreneurial, and tech elite. Many of the architectural wonders of Detroit's industrial past are being converted into innovation labs, workspaces, and playgrounds for Detroit's young adventurers. A vibrant downtown coworking space called the Green Garage was once a showroom for Model T automobiles in the 1920s. TechTown, a Detroit business incubator, was the former factory where General Motors designed the Corvette. The M@dison Building, once the opulent Madison Theater, is now a five-story, fifty-thousand-square-foot space for technology start-ups.

The creators of Ponyride approached Detroit's foreclosure crisis differently: they saw the space and the culture they could create as an asset for the community rather than an asset for an elite few. The two-story thirty-thousand-square-foot industrial building that houses Ponyride was an opportunity to conduct, in the words of management, "a study to see how the foreclosure crisis can have a positive impact on our communities." The creators of Ponyride were not referring to an academic study but rather to a real-world experiment in entrepreneurship, community building, and empowerment.

The enterprises that have called Ponyride home represent the remarkable diversity of ideas and innovators that populate the neighborhoods of Detroit. They include, for example, filmmakers, small industrialists, fashion designers, dancers, nonprofits, food and beverage sector businesses, furniture makers, printers, eyeware salespeople, and health and beauty product experts. As I studied the Ponyride model, several of the projects stood out for their ingenuity and community impact. Consider, for example, Detroit SOUP, a community-driven micro-funding program; and Empowerment Plan, an organization that hires and trains homeless women to make coats that convert into sleeping bags. These enterprises represent the diversity and vitality that characterize the Ponyride model. They also illustrate how the innovation ecosystem created by Ponyride is spreading out to include more and more of the larger Detroit community.

A DEMOCRATIC EXPERIMENT IN MICRO-FUNDING

Detroit SOUP is a micro-funding project designed to build stronger communities through innovation, collaboration, and inspiration. For a five-dollar donation, a visitor gets soup, salad, bread, and a vote. First, visitors listen as four people present four-minute pitches that explain their ideas for social enterprise. Over the years the pitches have focused on art, social justice, urban agriculture, and social entrepreneurship. After the presentations, visitors get their food and discuss the pitches they have heard. At the end of the evening, attendees vote on the pitch they think was the best. The presenter with the most votes takes home all the donations.

The first SOUP dinner was held on Super Bowl Sunday in 2010. No one pitched a project that day. Today hundreds of people submit projects to be pitched, creating a great deal of competition simply to present at a SOUP event. On average, roughly three hundred people attend a SOUP event, underscoring just how popular the community-based micro-funding program has become. According to SOUP, more than 25,000 people have attended the dinners since that first event in 2010. The projects have also raised more than $132,000 for Detroit innovators, many of them young creatives looking to make a unique impact in the city.

SOUP projects have varied greatly, highlighting the creativity that lives in Detroit's neighborhoods. One project called "Sit On It Detroit" repurposes wood from the tens of thousands of abandoned homes in the city and custom-builds benches for people to sit on while waiting for public transportation. The two young creatives behind Sit On It Detroit also decided to use the bottom half of the bench to store books that residents could read while waiting for their buses to arrive. Another project, "Kid's Summer Adventure," focuses on children from first through eighth grades and works to introduce them to nutrition literacy. In an effort to curb childhood obesity in the East Jefferson neighborhood, the program teaches kids about the benefits of a healthy diet, cooking, gardening, and exercise. A third project, "Shakespeare in Detroit," uses local talent—artists, designers, and actors—to perform Shakespeare in the places that local Detroiters live, work, and play, and brings both energy and theater to residents and their communities.

SOUP projects reflect a wide range of ideas, but they have a few things in common. First, they are all designed to improve the quality of life in Detroit communities. Additionally, the projects typically turn Detroit liabilities, such as the many abandoned building and public spaces, into assets.

The amount of money generated by the micro-funding program operated by SOUP is impressive. But the impact of SOUP is far greater than the money that has been distributed to the presenters over the years. SOUP stimulates the cultivation of human and social capital and thus is part of the new innovation economy. Even the projects that do not receive the prize at the end of the evening get invaluable feedback from local artists, entrepreneurs, and social innovators who attend SOUP events. These conversations invariably help give a novel idea more clarity and purpose in its mission or execution and also provide young creatives with a growing network of potential mentors, sponsors, collaborators, and funders.

"WE DON'T NEED COATS, WE NEED JOBS!"

Veronika Scott was attending art college in Detroit, studying product design. Like most design students, she expected to make shiny objects such as dishwashers and computers for a living after graduation. But that all changed one day when a humanitarian group visited her class and issued an assignment: design something to fill a need in the city. Scott figured that in a hardscrabble place like Detroit, it would not be difficult to "find people with bigger needs" than her own. She decided to visit a homeless shelter for inspiration.

Scott began talking with the homeless people there, and together they came up with an idea—design a coat that could convert into a sleeping bag during the freezing Detroit winters. The first coat, in Scott's words, "looked like a body bag." The design was horrible, but the idea was laudable. "I spent eighty hours to sew trash bags and seven wool coats to build a Frankenstein twenty pound coat tent," she recalled. But she persisted, and the design improved. Scott visited homeless shelters three times a week for five months, soliciting ideas from members of the homeless population and learning more about their personal experiences. When the school year ended, she continued visiting shelters and speaking with homeless Detroiters. The project and the people she had met seemed too important to abandon simply because she had completed a class assignment.

As she was leaving a shelter one day, a woman followed her outside. "We don't need coats, we need jobs!" the homeless woman told Scott. The insight was pivotal and from a perspective that only a person in the homeless woman's position could offer. Scott was a designer by training. Up to this point, her mission had been focused on the product—the coats. But as a result of what her homeless acquaintance had said, Scott's vision and mission changed. She became a young creative focused on social entrepreneurship. The strategic pivot shifted her goal from making coats to remaking the lives of homeless women she encountered through employment.

Many of the women had ended up homeless after leaving abusive relationships to protect themselves and their children. The coats were a good idea, but they were also a short-term solution to a more endemic problem—joblessness, a cycle of poverty, and a lack of financial independence. Helping homeless women to develop job skills and become self-sufficient was a long-term and truly transformational solution. This became the core mission of a nonprofit that Scott created called the Empowerment Plan.

When I met with Hailey McInnis, the operations manager of the Empowerment Plan, she was excited to share how the organization's vision was continuing to evolve. "We are busting out of the seams in our space in Ponyride," McInnis told me when we spoke. The organization had become so successful that they were moving into their own building and hiring more seamstresses, as well as women to work on their retail team. In 2017 they were scheduled to distribute more than 6,500 coats in various markets, including Detroit, Chicago, Boston, San Francisco, and the state of Utah. Their clients included homeless shelters across the US. They also supplied coats to countries as far away as New Zealand, Belgium, and Switzerland. The Empowerment Plan maintained strategic partnerships with corporations like General Motors and professional hockey teams like the Detroit Red Wings and football teams like the Detroit Lions and the New England Patriots. The pop star Madonna, a Michigan native, donated funds to support hiring and educating two women from a homeless shelter.

"Our goal has really shifted from the coats to employment," McInnis said. "We needed more space to accommodate the increased number of women we plan to hire." When McInnis and I spoke, the Empowerment Plan employed twenty-four homeless women. In the new space, they would be able to increase that to sixty, in addition to the retail team they were building.

McInnis explained that the organization was devising a "new one- to three-year model." The idea is to help women transition from working as seamstresses to other lines of work, if that is something they desire. To do this, the Empowerment Plan was devising a well-crafted education program that was becoming just as significant as the production of the coats. The Empowerment Plan requires every homeless woman they hire to have at least a GED. If they do not have a GED, the program pays for them to take classes and earn the credential. Importantly, the Empowerment Plan pays women while they are in classes, thus making sure that they are able to earn full-time wages.

I was struck by Scott's approach to innovation. Whereas Ponyride requires its members to offer education as a service, receiving education was a built-in feature of the model established by the Empowerment Plan. The young creative's idea to transform the lives of homeless women not through charity but through employment is a brilliant reflection of the new innovation economy. Moreover, the success of the Empowerment Plan speaks powerfully to how Ponyride has cultivated an ecosystem in which young creatives are free to fully recalibrate what innovation looks like, what innovation does, and whom it serves.

In some respects the Ponyride model has been too successful. The combination of designers, civic entrepreneurs, artists, and small-scale manufacturing contained within its walls has put the coworking space out of compliance with the city's building codes. In order to become compliant, Ponyride would have to make substantial upgrades to the facility, which would have been expensive. Rather than pass that cost on to the tenants and, in effect, take away one of the premier benefits of Ponyride membership—affordable rent—management decided to sell the building and move into a new space, one shared with Detroit's Make Art Work, which is housed in a 160,000-square-foot complex for artists and community organizations. The $100,000 investment that Cooley made in Ponyride back in 2009 had increased in value to an estimated $3 million in 2018. This rising value was a reflection of the rising value of property in the Corktown neighborhood as well as the hard work the Ponyride team had put into reinventing the facility.

WHY DETROIT MATTERS IN THE
NEXT INNOVATION ECONOMY

After decades of population losses, industrial erosion, and economic decline, the resurgence of innovation in Detroit is instructive. The city offers us a critical glimpse into the future, not because of the robust downtown economy that has emerged but rather because of the Ponyride experiment. Ponyride illustrates the fertile, sustainable aspects of an innovation ecosystem that is accessible to the many rather than the few. Ponyride also expands our notions of what the innovation enterprise can create.

Every city should have a Ponyride—a space, that is, that empowers community members to participate in the innovation enterprise. One radical idea would be to bring the Ponyride model into public schools, making them hubs of design, collaboration, and innovation. Schools are largely out of touch with the real world and the kinds of skills that young people should be cultivating. Including spaces like Ponyride in schools would encourage students to act as agents in their own learning rather than as receptacles of ideas chosen by someone else. Schools would instantly be transformed into spaces where young people participate in experiential learning, designing and creating artifacts, ideas, and social and entrepreneurial initiatives that impact the lives and communities around them. Imagine students regularly having the opportunity at school to feel the childhood sensation of getting on a pony, falling off, and getting back on to try again. In other words, imagine students having the opportunity to become designers, problem-solvers, entrepreneurs, and innovators. Unfortunately, the current batch of educators, policy makers, and legislators seem woefully incapable or unwilling to build a more expansive and relevant vision of education.

Even without the investments and partnerships that are driving the growth of Detroit's downtown economy, young creatives across the city have mined their own resources to catalyze a community-based approach to innovation that includes the low-opportunity neighborhoods where so many Black and poor residents live. If the tech, economic, and political elite have neglected the entrepreneurial and civic talent in the city's neighborhoods, young creatives have taken full advantage of resources like Ponyride to expand the innovation enterprise to a very different cast of characters.

THE MEANING OF YOUNG CREATIVES

Millennials are a much-maligned generation. In a world they inhabit but didn't create, they are frequently castigated for speaking out about such issues as the college debt crisis, the disappearing middle class, and the erosion of opportunity. And even as the economy has sputtered toward recovery after the Great Recession, young people must contend with a "new normal" marked, most notably, by the likelihood of being the first generation in memory to not surpass their parents' economic standard of living.

In response to the world they have inherited, young creatives are mobilizing a very distinct set of assets—their intellect, technology, vibrant social networks, and resilience—to rethink and remake key institutions, including education, tech, media, entertainment, and civic life. Like Prince Harvey, the rapper who turned an Apple Store into his studio, young creatives do not need a lot. Like Issa Rae, the awkward Black girl who leveraged the few resources she had to reimagine television, young creatives are gathering the few resources they have and using them to pursue the lives that they want. Like Wiley Wiggins, the young UX designer who only had a high school diploma, young creatives are discovering new ways to learn the things that our schools are not teaching them. As Alexandria Ocasio-Cortez suggests, young creatives recognize that rather than knocking on the door to someone else's house, you might as well build your own.

Rather than dismiss millennials as entitled, tech-addicted, narcissistic, and ill-prepared for the challenges of adulthood, their older counterparts will be better served appreciating and replicating the innovation economy that young creatives are building—because it serves us all. Through sheer hustle and imagination, young creatives are building an economy the likes of which we have never seen before. They are designing solutions that open up the terrain of innovation to new people, new places, and new possibilities. They are privileging principles like diversity, inclusion, and social impact over homogeneity, exclusion, and personal wealth. In a world marked by a widening wealth gap, faltering schools, and rising cultural and political resistance to change, millennials are staging the ultimate cultural disruption by expanding who practices innovation and the very geography of innovation.

Let's join them and figure out ways to support their efforts to reinvent the innovation economy. Or let's at least get out of their way. They may save us all.

ACKNOWLEDGMENTS

This book is the by-product of many generous people who offered me their time and expertise while I was researching and writing. The idea for the book was inspired by my work with the Connected Learning Research Network, a group of scholars funded by the MacArthur Foundation—more specifically, numerous meetings and conversations with Mimi Ito, Sonia Livingstone, Kris Gutiérrez, Juliet Schor, Jean Rhodes, Bill Penuel, Dalton Conley, Vera Michalchik, Julian Sefton-Green, and Ben Kirshner.

The research team that I was able to coordinate to conduct the field-work—observations, interviews, filming, data analysis—was made possible by the MacArthur Foundation and its support of the Digital Media and Learning initiative. Several people at the MacArthur Foundation made this work possible, including Julia Stasch, president, and Jennifer Humke, senior program officer. Connie Yowell, the former director of education at MacArthur, was a critical catalyst for the investments MacArthur made to better understand young people's engagement with digital technologies.

The fieldwork that I did for years was greatly enhanced by the contributions of top-notch PhD, MA, and MFA students. Andres Lombana-Bermudez, Alexander Cho, Krishnan Vasudevan, and Robyn Keith brought the gaze of social science and critical thinking to the project. They did ethnographic research in a variety of places, including coworking spaces, game jams, meetups, open mic nights, and hackathons. Robin McDowell brought her expertise in art and design to develop important strategies for communicating our research and developing a web-based presence, doinginnovation .org. Monique Walton shared her expertise in independent filmmaking to help capture a lot of our research via video and short stories. Monique and Krishnan produced media that helped to crystallize our storytelling and analysis.

Beacon Press continues to help support my research and my desire to share it with a broader audience. I am especially grateful to Gayatri Patnaik, whom I have worked with on two previous books, for introducing me to my editor for this project, Rakia Clark. From the very beginning Rakia saw the potential for this book and was a champion for it all the way through.

I'd also like to thank my colleagues from the University of Texas, including Sharon Strover, Tom Schatz, Alisa Perren, Wenhong Chen, Kathleen Tyner, Joseph Straubhaar, and Sherri Greenberg for helping me to sharpen my understanding of media, technology, culture, innovation, and society.

Parts of this work have been shared at many universities and organizations over the years, eliciting both feedback and support that has been especially generous. Many organizations and people opened their doors to me and my team, which enriched our insights and informed my writing. I am especially grateful for the many people that I profile in this book who gave of their time to be interviewed or observed.

Finally, I'd like to thank my amazing wife, Angela Hall Watkins, and my daughter, Cameron Watkins, for nourishing me with love, laughter, and life. It was not easy to accommodate me as I was essentially working on two book projects simultaneously, but they did so with sincerity and understanding. Thank you.

NOTES

Introduction: Respect the Hustle

1 "We were just like, 'Don't check it . . .' ": Newman, Wang, and Ferré-Sadurni (2018).

2 "Lots of these folks were mad . . .": Gonzalez-Ramirez (2018).

2 "In my opinion, if women . . .": Gonzalez-Ramirez (2018).

2 "I just started running . . .": Newman, Wang, and Ferré-Sadurni (2018).

2 "the most significant loss for a Democratic incumbent . . .": Goldmacher and Martin (2018).

2 The *Washington Post* called it the "defining upset" . . . : Bump (2018).

3 "New York Fourteen is the source of a lot . . .": Ocasio-Cortez, "Alexandria Ocasio-Cortez Tackles" (2018).

3 "I come from a working-class background . . .": Chávez and Grim (2018).

3 "A lot of people of color were excited . . .": Goldmacher (2018).

3 Steven Romalewski, director of the Mapping Service . . . : Jilani and Grim (2018).

5 "I started this race, nine, ten months ago . . .": Ocasio-Cortez, "Political Newcomer" (2018).

6 "I felt like at this point . . .": Ocasio-Cortez, "Political Newcomer" (2018).

6 According to the Bureau of Labor Statistics . . . : US Bureau of Labor Statistics, "Contingent and Alternative Employment" (2017).

7 "I understand the pain of working-class Americans . . .": Ocasio-Cortez, "Political Newcomer" (2018).

9 "unions, the working class, progressive candidates . . .": Daalder (2018).

10 "to sell Americans on socialism" . . . : Beer (2018).

10 "It had a clear leftist vision that . . .": Jilani (2018).

10 "a very small footprint production . . .": Daalder (2018).

10 "a ragtag group of filmmakers . . .": Jilani (2018).

11 In 2015 they became the largest . . . : Fry, "Millennials Are the Largest Generation" (2018).

11   Adjusted for inflation, the median earnings . . . : US Census Bureau, "Young Adults" (2015).

11   It is also an age in which . . . : Chetty et al. (2016).

12   the "hustler creative" . . . : Quinn (forthcoming).

12   "I'm part of this work generation . . .": Sow (2017).

12   A 2016 study by the National Bureau of Economic Research . . . : Katz and Krueger (2016).

13   The US Bureau of Labor Statistics reports that . . . : US Bureau of Labor Statistics, "Contingent and Alternative Employment" (2018).

Chapter 1: "You Don't Need a Lot"

15   "It wasn't my plan to record this . . .": Narvin (2015).

16   "People would also ask me for tips . . .": Beaumont-Thomas (2015).

16   "Sometimes I would get a little loud . . .": Beaumont-Thomas (2015).

16   He called the effort "a cappella" hip hop...: Prince Harvey, Facebook, February 22, 2016, https://www.facebook.com/princeharveylove/videos/i-recorded-my-album-at-an-apple-store-this-is-my-first-music-/1693077680946260.

16   "A lot of people give you excuses . . .": Barnes (2015).

20   The side-hustling economy is usually . . . : For more on young creative types and the nighttime economy, see, for example, Currid (2008) and Neff (2012).

20   "There is no leeway for chancy trial, error . . .": Jacobs (2011, 245).

20   "New ideas need old buildings": Jacobs (2011, 245).

21   According to a report produced by . . . : Katz and Krueger (2016).

21   "This emerging occupational landscape . . .": For some historical perspective on the transition from a goods-producing economy to a service-oriented economy, see Bell (1973) and David Harvey (1990).

21   By 2022 that figure is expected . . . : McBride 2017).

22   "One early study from the University of Texas . . .": Spinuzzi (2012).

22   "Connections with others are a big reason . . .": Spreitzer, Bacevice, and Garrett (2015).

23   A study by the US Bureau of Labor Statistics found that...: US Bureau of Labor Statistics, "Contingent and Alternative Employment" (2018).

23   "working together to make the world . . .": The descriptions and analysis of the Chicon coworking space are based on scheduled interviews and observations conducted by Robyn Keith, a PhD student in sociology, as well as my many visits to, observations of, and interviews with members of Chicon. The bulk of these observations were conducted in the summer of 2014, with some follow-up observations in 2015.

29   They even received a favorable write-up . . . : Horgan (2014).

31   Today the complex is called 501 Studios . . . : The descriptions and analysis of the activities at the North Door are based on scheduled interviews and observations conducted by Andres Lombana-Bermudez, a PhD student in communication. The author also conducted some observations and interviews with some of the participants in the Juegos Rancheros indie game collective. The bulk of these observations were conducted in the summer of 2014, with some follow-up observations in 2015.

32   The meetups established the environment . . . : For more on the concept of open innovation, see, for example, Chesbrough (2003).

33   A report by the Brookings Institution . . . : Katz and Wagner (2014).

33   These companies represent what . . . : Sundararajan (2017).

Chapter 2: Bootstrapping

37   The city has long maintained a reputation as . . . : Moretti (2012).

38   Even as Austin was adding educated . . . : For more on key shifts in Austin's population, see, for example, Tang and Ren (2014).

38   A *New York Times* exposé in 2015 . . . : Kantor and Streitfeld (2015).

39   The IGDA concludes that the prototypical . . . : Weststar, O'Meara, and Legault (2018).

39   People in the industry call this "crunch" . . . : For more on the practice of "crunch" in the game industry, see, for example, Schreier (2017).

39   More than half, 51 percent . . . : Weststar, O'Meara, and Legault (2018).

39   It also explained that permanent . . . : Weststar, O'Meara, and Legault (2018, 19).

39   "cognitive stratification" . . . : Crawford (2009).

42   The Long Tail theory suggests that as . . . : Anderson (2006).

44   In 2017 half of the eighteen top-subscribed-to . . . : McAlone (2017).

45   More than 700 million people . . . : Alexander (2010).

46   According to the Entertainment Software Association . . . : Entertainment Software Association (2013).

49   The main point is to generate . . . : For a discussion of rapid prototyping, see, for example, Brown (2009).

49   "the smartphone's absence of buttons . . .": Parkin (2013).

49   *"Canabalt's* solution was elegant and simple . . .": Parkin (2013).

51   "Things look much rosier from . . .": Rao (2018).

51   A 2017 report by the IDGA . . . : Weststar, O'Meara, and Legault, 2018.

52   A 2016 report released by CareerBuilder . . . : CareerBuilder (2016).

53   Further, in a study of two innovation hubs . . . : Richard Florida has produced a series of reports and articles that examine the concentration of venture capital in a small number of geographic areas. See, for example, Florida, "The Amazon Model" (2013), and Florida and King (2016).

54   Social scientists call this the "herding effect" . . . : For examples of the herding effect in crowdfunding, see Bretschneider, Knaub, and Wieck (2014). Kuppuswamy and Bayus (2013) argue that herding behavior can negatively affect a crowdfunding project.

54   Foster explained that close friends and family . . . : Granovetter (1973).

55   "Games have seeped into every aspect . . .": Bogost (2011, 11).

56   Scholars have used terms . . . : For a discussion of "venture labor" in the digital economy see, for example, Neff (2012). For a discussion of "aspirational labor," see, for example, Duffy (2017). For a discussion of "hope labor," see, for example, Kuehn and Corrigan (2013).

Chapter 3: The School of the Internet

57   In 2016 researchers from Stanford University's . . . : Chetty et al. (2016).

57   Examining US Department of Labor data . . . : Fry (2017).

58   In one study a team of economists . . . : Gervais et al. 2014).

58   "Young workers . . . spend more time . . .": Gervais et al. 2014, 54).

58   "People who switch jobs more . . .": Thompson 2014).

58   "Young people aren't quitting more . . .": Thompson, 2014.

59   This reflects the rise of what some call . . . : For more on the risk economy, see, for example, Hacker (2006).

59   As the economy continues trending toward . . . : For more on skill-biased technical change, see, for example, Acemoglu and Autor (2011).

65   This is especially true for students . . . : Watkins et al. (2018).

66   "Why do English classes focus . . .": Caplan, 2018.

66   "The labor market doesn't pay you . . .": Caplan (2018).

67   A study by the Pew Research Center . . . : Taylor, Fry, and Oates (2014).

67   In 1965 the average income gap . . . : Taylor, Fry, and Oates (2014).

Chapter 4: Hustle and Post

71   "We were at the university . . .": Ljung (2011).

71   "I made some terrible music, but . . .": Ljung (2011).

71   "Everybody doing music on the web . . .": Ljung (2011).

71   "It was just really, really annoying for us . . .": Van Buskirk (2009).

71   "I mean simple collaboration, just sending . . .": Van Buskirk (2009).

71   "We didn't have that kind of platform . . .": Van Buskirk (2009).

71   "We decided to focus on the creator side . . .": Ljung (2011)

74   "the most vital and disruptive new . . .": Caramanica (2017).

75   "tightly penned raps, sonic cohesion . . .": Nosnitsky (2011).

75   "may just be the future of the music business . . .": Nosnitsky (2011).

76   "Everything is literally made with me . . .": Caramanica (2017).

77   In a 2017 interview with the *New York Times* . . . : Caramanica (2017).

77   "This generation and wave of rap . . .": Turner (2017).

77   "Mainstream acts are only going to make . . .": Turner (2017).

77   But young people also use social media . . ." The *Millennials, Social Media, and Politics* survey (Rideout and Watkins [2019]) was conducted September 27–October 21, 2017, using the probability-based AmeriSpeak® Panel from NORC/University of Chicago. The survey was offered via the web in English and Spanish. The survey results included 1,010 respondents ages eighteen to thirty years old in the United States. It was designed by S. Craig Watkins and Victoria Rideout.

78   Hip-hop creatives saw the future of music . . . : For a discussion of rap music's early embrace of the internet, see Watkins (2005).

78   In 2012 hip-hop icon Dr. Dre . . . : Sisario (2014).

78   "from a pure sales-based ranking . . .": Billboard Staff (2014).

78   The algorithm, developed by Nielsen Entertainment . . . : Billboard Staff (2014).

78   The *Billboard* charts and the business of pop music . . . : For a more detailed discussion of SoundScan's impact on the business of pop music, see Watkins (2005).

79   By contrast, physical sales (i.e., CDs, vinyl) accounted for . . . : All of the reported revenue data for recorded music is from Friedlander (2018).

79   In 2017 Lil Pump's single "Gucci Gang" . . . : Suarez (2017).

79   Lil Uzi Vert's album *Luv Is Rage 2* . . . : Coscarelli (2017).

79   In 2018 SoundCloud rapper XXXTentacion's album . . . : Zellner (2018).

79   Another SoundCloud rapper, 6ix9ine . . . : Caulfield (2018).

79   Shortly after XXXTentacion was shot dead . . . : Sisario and Coscarelli (2018).

79   The success of SoundCloud rap precipitated . . . : Knopper (2018).

80   The arrival of a new CEO . . . : Satariano and Shaw (2017).

81   "artists were always SoundCloud's core value . . .": Nicolaou (2018).

Chapter 5: The People's Channel

83   "That's one of those things about being . . .": Karim (2007).

84   "Our little website was finally taking off . . .": Karim (2007).

84   Fans of the show eventually revealed . . . : Heffernan and Zeller (2006).

85   "I was just working from job to job . . .": Rae, "Issa Rae, Creator" (2013).

86   "wasn't going like I really hoped" . . . : Rae, "Awkward Black Girl: A Conversation" (2016). Some of this chapter is also informed by Rae's memoir *The Misadventures of Awkward Black Girl.*

86   "I knew I was black for a long time . . .": Rae, "Our Un-Awkward Conversation" (2012).

86   "I was really struggling . . .": Rae, "Issa Rae, Creator" (2013).

86   "Where's the Black Version of Liz Lemon?": Pitterson (2010).

86   "I think my little sister, and all the other . . .": Pitterson (2010).

87   "I owned a camera, I owned a computer, . . .": Rae, "HAATBP Keynote" (2014).

87   "I tried to pitch one of my other web series . . .": Rae, "Issa Rae of *Awkward Black Girl*" (2012).

87   "The internet," she would say later . . . : de Cadenet (2013).

89   When she and a colleague are teamed together . . . : *The Misadventures of Awkward Black Girl,* season 1, episode 9, "The Happy Hour," directed by Issa Rae and Shea Vanderpoort, written by Amy J. Aniobi and O. C. Smith, posted October 5, 2011.

90   "It was like watching something about me . . .": Pye, Tiffany. "I'm so glad to have supported such an amazing project." Comments. *The Misadventures of Awkward Black Girl.* Kickstarter. https://www.kickstarter.com/projects/1996857943/the-misadventures-of-awkward-black-girl/comments.

90   "I appreciate what your show means . . .": Redhead, Bianca. "Im so excited to see another awkward black female stanford grad making important strides." Comments. *The Misadventures of Awkward Black Girl.* Kickstarter. https://www.kickstarter.com/projects/1996857943/the-misadventures-of-awkward-black-girl/comments.

90   Rae told *Fast Company* in 2012 . . . : Berkowitz (2012).

91   Live tweeting functions as a . . . : Kimble (2017).

92   "It's so hard to have a nine to five . . .": Eler (2016).

93   "Television today has a very limited . . .": Rae, *The Misadventures* Kickstarter (2011).

93   "help us change the face of Hollywood." . . . : Rae, *The Misadventures* Kickstarter (2011).

94   Almost half, 46 percent, of the backers . . . : Rae, 2011. The analysis of the supporters of the *Awkward Black Girl* Kickstarter campaign is based on the Community data provided by the crowdfunding campaign.

94   "I started Color Creative to give opportunities . . .": Naasel (2014).

94   She envisioned Color Creative as . . . : Naasel (2014).

95   The study, titled *Race in the Writers' Room* . . . : Darnell Hunt (2017).

95   "I was trying to break into the industry . . .": Peoples (2016).

96   In January 2018 YouTube announced . . . : Mohan and Kyncl (2018).

Chapter 6: Can You Hear Us Now?

98   A 2017 report by the Directors Guild of America . . . : Directors Guild of America (2017).

98   African Americans and Latinos make up 4 and 6 percent . . . : Hayes and Chmlielewski (2018).

98   *Bloomberg* reported that just "one of the all male executives . . .": Shaw (2018).

99   "I remember I was watching TV and all of a sudden . . .": Simien, *"Dear White People* Director" (2014).

99   "a Benetton ad," the "white as snow . . .": Simien, *"Dear White People* Is a Satire" (2014).

99   "It wasn't even that it was white . . .": Simien, "Justin Simien's Open Letter" (2014).

99   Life at Chapman was a constant source . . . : The details about Justin Simien's life and work in this chapter come from interviews by Hillary Sawchuk (Simien, *"Dear White People* Director," 2014); Terry Gross (Simien, *"Dear White People* Is a Satire," 2014); and Chase Jarvis (Simien, "How to Overcome NO," 2014).

100   "love and theft" legacy of blackface minstrelsy . . .": Lott (1993).

100   He told himself, "I need to do this . . .": Simien, "How to Overcome NO" (2014).

100   "Whether the script gets sold . . . ": Simien, "How to Overcome NO" (2014).

100   "feeding the creative beast" . . . : Simien, "How to Overcome NO" (2014).

101   "To get the film done and be able to pay my rent" . . . : Simien, "How to Overcome NO" (2014).

102   "Let's make a trailer treating the movie . . .": Simien, "How to Overcome NO" (2014).

103   "the sort of backyard concept trailer . . .": Simien, "How to Overcome NO" (2014).

103   "Instead of using the concept trailer . . .": Simien, "How to Overcome NO" (2014).

103   "Remember when Black movies didn't necessarily star . . .": Simien, *Dear White People*, Indiegogo (2012).

104   "Hollywood is a place . . .": Gabbatt (2014).

104   "Hey, this great movie is coming . . .": Simien, How to Overcome NO" (2014).

105   "I'm trying to bring a new story . . ." Simien, "The Man Behind" (2012).

105   "My film really isn't about 'white racism . . .'" and "It's about the difference between . . .": Simien, "The Birth of *Dear White People*" (2012).

105   Hollywood's lack of interest in films that probe the complexities . . . : For more on Spike Lee's combination of art-house cinema, Black representational politics, and commercialism, see Watkins (1998).

105   The crowdfunding campaign was a success . . . : Simien, "The Birth of *Dear White People*" (2012).

105   "Black audiences are thought of . . ." and "People were excited that it was . . .": Berstein (2013).

106   "I had been working on the screenplay . . .": Simien, *"Dear White People* Is a Satire" (2014).

106   "a way to see which of Sam's jokes . . .": Barone (2014).

107   "When you walk into a room . . .": Rose (2015).

107   "Films with predominantly white casts . . ." and "For black films, you have the . . . ": Hill (2013).

108   After investigating more than . . . : Smith, Choueiti, and Pieper (2017).

108   Seventy percent of the characters in film . . . : Smith, Choueiti, and Pieper (2017).

109   "It really, really helped that I never . . .": Simien, "How to Overcome NO" (2014).

110   "If it weren't for social media . . .": Rae, "HAATB Keynote" (2014).

110   "Online content and new media . . .": Rae, "HAATB Keynote" (2014).

Chapter 7: STEM Girls

111   Whereas the projected growth rate for all occupations . . . : For a more detailed explanation of these rates, see US Bureau of Labor Statistics, *Employment Projections— 2016–2026* (2018).

111   Already, STEM workers, on average, earn 26 percent . . . : Langdon et al. (2011).

112   In 2014, 70 percent of all Google employees . . . : Google (2018). For news reporting on Google's release of data in 2014, see Gibbs (2014) and Miller (2014).

112   Facebook's workforce that year was 69 percent . . . : Williams (2014).

112   Yahoo reported that men made up 62 percent . . . : Reses (2014).

112   Microsoft's workforce was 72 percent male . . . : Endler (2014).

112   Unconscious bias, in this context . . . : For a more in-depth explanation of unconscious bias, see Greenwald and Banaji (1995).

113   In 2016 Google and Facebook launched . . . : Vara (2016).

113   One of the main pathways to a successful career in STEM . . . : All of the STEM bachelor's degree attainment data in this paragraph is detailed in Musu-Gillette et al., "Indicator 24: STEM Degrees" (2017).

113   There is gender variation in STEM degree . . . : All of the data cited in this paragraph is from Snyder, de Brey, and Dillow, "Degrees Conferred By Postsecondary Institutions" (2018).

114   In 2013 and 2014 women earned 35 percent . . . : All of the data cited in this paragraph is from Musu-Gillette et al., "Indicator 24: STEM Degrees" (2017).

114   He argues that skills gained . . . : Heckman (2000).

114   According to the National Center for Education Statistics . . . : The data cited in this paragraph is from Musu-Gillette et al., "Indicator 10: Mathematics Achievement" (2017).

115   The gap in math test scores between male and female . . . : Musu-Gillette et al., "Indicator 10: Mathematics Achievement" (2017).

115   "The argument continues to be made . . .": Florida State University (2017).

115   "But when we hold mathematics ability . . .": Florida State University (2017).

115   Perez-Felkner's research team found . . . : Perez-Felkner, Nix, and Thomas (2017).

116   In 1950 women accounted for . . . : Toossi and Morisi (2017).

116   The Bureau of Labor Statistics reports that in 2016 . . . : US Bureau of Labor Statistics, *Highlights of Women's Earnings* (2017).

116   For example, women in engineering occupations *and* A female software developer earns . . . : US Bureau of Labor Statistics, "Median Usual Weekly Earnings," 2017.

117   "Why would a creative, artistic girl . . .": Sterling, "Yesterday's Frontiers," 2013.

117   It was not surprising that Sterling . . . : In 2017 women made up 16 percent of engineers in the US: US Bureau of Labor Statistics, "Labor Force Statistics," 2018.

117   "It was so cool, and so much fun . . .": Sterling, "Yesterday's Frontiers" (2013).

117   "In that class I learned that engineering . . .": Sterling, "Yesterday's Frontiers" (2013).

118   "Belonging," Dasgupta told the National Science Foundation . . . : Dasgupta (2016).

118   "Belonging is just a way of saying . . .": Dasgupta (2016).

118   "feeling like they fit in . . .": Dasgupta (2016).

118   "Great," she says . . . : The quotes in this and the following paragraph are from Sterling, "Yesterday's Frontiers" (2013).

119   "Once a month, we'd get together . . .": Sterling, "How the Founder" (2017).

119   "My friend, Christy, . . . got up and started . . .": Sterling, "How the Founder" (2017).

121   "What if I put those two things . . .": Sterling, "Yesterday's Frontier" (2013).

122   "What I encountered instead . . .": Sterling, "Yesterday's Frontier" (2013).

123   In less than two years Sterling . . . : The GoldieBlox Super Bowl commercial in 2014 was the result after Sterling won a contest conducted by Intuit to support small business. See Kavilanz (2014).

123   "the iconic retail chain . . .": Corkery (2018).

123   In 2018 a series of reports in the *New York Times* . . . : Bowles, "Dark Consensus" (2018); "Digital Gap" (2018).

124   There is an emerging body of research . . . : For more on co-viewing research, see Valkenburg et al. (1999) and Takeuchi et al. (2011).

124   "We've launched multiple YouTube series . . .": Sterling, "How the Founder" (2017).

124   "to actually result in learning . . .": Elizabeth Foster, "GoldieBlox" (2016).

124   The GoldieBlox YouTube channel reflects . . . : Ault (2014).

124   YouTube has come under fire for . . . : Maheshwari, "On YouTube Kids" (2017).

125   "Silicon Valley now has a responsibility . . .": Maheshwari, "YouTube Is Improperly Collecting" (2018).

Chapter 8: Code for Change

127   For example, judges across the US . . . : Israni (2017).

128   Politicians and political consultants are deploying . . . : Much has been made of the way two high-profile presidential campaigns, Obama 2012 and Trump 2016, used social media and data to micro-target voters. See Urbain (2018) and Hoover (2018).

129   In 2017 those who took the exam . . . : The data cited in this paragraph was compiled by Ericson (2018).

129   According to the 2015 National Assessment of Educational Progress's . . . : von Zastrow (2016). Also see Chang (2016).

129   A 2016 study by Google and Gallup . . . : Google Inc. & Gallup Inc. (2016).

130   In my own research I have observed . . . : As part of a research network funded by the MacArthur Foundation, I led a team of advanced graduate students to study some of the notable ways social and educational inequality persists in the US despite the widespread adoption of the internet. We write extensively about the findings in our book *The Digital Edge: How Black and Latino Youth Navigate Digital Inequality* (Watkins et al., 2018).

130   The Bureau of Labor Statistics projects that employment . . . : All of the data cited in this paragraph is from US Bureau of Labor Statistics, "Computer and Information" (2018).

130   In the US some states have even . . . : Zubrzycki (2016).

131   "Few of my classmates looked . . .": Black Girls Code (2018).

131   "While we shared similar aspirations . . .": Black Girls Code (2018).

132   In 2016 women received 57 percent . . . : National Center for Women and Information Technology (2017).

132   Among software developers, white women . . . : The data on the percentage of Latinas, white, and Black women working as computer programmers and software developers is compiled by Ashcraft, McLain, and Eger (2016).

132   "Girls of color can learn computer science . . .": Westervelt (2015).

133   In 1985 women earned 37 percent . . . : Snyder and Dillow (2013).

133   A 2017 study funded by Microsoft . . . : Kesar (2018).

133   Kesar's survey finds that female . . . : Data from this paragraph is detailed in Kesar (2018, p.7).

135   "The volume of evidence shows . . .": Hickey (2018).

135   In Shalini Kesar's study . . . : Data from this paragraph is detailed in Kesar (2018, 8).

136   She reports that girls in grades five through twelve . . . : Data from this paragraph is detailed in Kesar (2018, 10).

136   Moreover, the study suggests that STEM clubs . . . : Data from this paragraph is detailed in Kesar (2018, 11).

136   Studies show that students who have . . . : Alexander, Entwisle, and Olson (2001).

137   This may explain the gap that has widened . . . : Kaushal, Magnuson, and Waldfogel (2011).

137   "the rug rat race." . . . : Ramey and Ramey (2010).

137   According to Shalini Kesar's study, in middle school . . . : Data from this paragraph is detailed in Kesar (2018, p. 6).

137   Researchers believe that girls are much less likely . . . : For more on the growth mind-set, see, for example, Dweck (2006).

137   Girls are much more likely than boys . . . : Perez-Felkner, Nix, and Thomas (2017).

138   The efforts of organizations like Black Girls Code . . . : The quote in the previous sentence and the data here is from Frey, "The US Will Become" (2018).

138   Today, most of the major metropolitan public school . . . : Musu-Gillette et al., "Indicator 6" (2017).

Chapter 9: Hacking While Black

141   In the late 1990s the US Department of Commerce . . . : National Telecommunications and Information Administration (1995); National Telecommunications and Information Administration (1998); National Telecommunications and Information Administration (1999).

142   In 1999, 74 percent of schools with . . . *and* By 2005 a Black ninth-grade . . . : Wells and Lewis (2006).

142   Over the years, countless studies have . . . : Hargittai (2002); Warschauer (2003); Hargittai and Hinnant (2008); van Dijk (2012).

142   Rather, as the technology has become . . . : Wei (2012); Pearce and Rice (2013).

142   This includes, for example, being able to . . . : For more on how computer-mediated social networks are used, see Rainie and Wellman (2012).

142   In fact, there is widespread recognition . . . : For an example of the access gap, see National Telecommunications and Information Administration (1999); for an

example of the skills gap, see Hargittai (2002); for an example of shifts in the access gap, skills gap, and participation gap, see Watkins et al. (2018).

143 In a series of reports for the *New York Times* . . . : Bowles, "Dark Consensus" (2018); "Digital Gap" (2018); "Silicon Valley" (2018).

144 For example, in 2015 about 35 percent . . . : Musu-Gillette et al. (2017).

144 Asian students, at 71 percent, are the most likely . . . : Musu-Gillette et al. (2017).

144 First, most young people will work an average of twelve to fifteen jobs . . . : Marker (2015).

145 Economists call this "skill-biased technical change" . . . : For more on skill-biased technological change, see Autor, Katz, and Krueger (1997).

145 "imply substantial workplace transformations . . .": Manyika et al., 2017.

146 Brynjolfsson and McAfee contend that ideation skills . . . : Brynjolfsson and McAfee (2014).

146 Rather than face a jobless future . . . : Kelly (2012).

147 "pervasive competency in the collaborative and iterative skills . . .": Pacione (2010).

147 Critics maintain that design thinking is really a clever . . . : Irani (2018).

147 Design thinking, from the view of critics . . . : Vinsel (2018).

149 "Hackathons are a chance for engineers . . .": Keyani (2012).

149 It was an internal hackathon at Facebook . . . : For inside accounts of how the hackathon culture led to the creation of Facebook's "Like" button, see, for example, Chan (2009) and Bosworth (2014).

149 "encourage students to tinker with new software . . .": Leckart (2015).

149 "The best talent are at these opt-in courses . . .": Leckart (2015).

150 "high-potential youth in low-opportunity settings" . . . : DeRuy (2015).

150 "Hackathons have been around for a while . . .": Cueva (2015).

150 "When I was eight years old . . .": Cueva (2015).

153 "To actually know that there are . . .": "Young Coder Finds Like Minds" (2014).

154 "help eliminate the feelings of danger . . .": Yahoo Finance (2016).

154 "It provides an easy gateway . . .": "Swimming in Their Genius" (2014).

155 "STOP," she told the audience, "is a platform . . .": Amis et al. (2015).

156 They are the catalyst behind the most powerful collective force . . . : For more on Black Twitter see, for example, Watkins et al. (2018).

156 By 2010 black and Latino teens' savvy use of . . . : For more on the distinct ways Black and Latino teens use digital media, see, for example, Watkins et al. (2018).

Chapter 10: Woke

159 "the chance to meet with young people . . .": Obama (2016).

159 "non-binary people . . . literally have . . .": Obama (2016).

159 "have made you change your mind . . .": Obama (2016).

160 "Once you've highlighted an issue . . .": Obama (2016).

160 "But," he said, "once people who are in power . . .": Obama (2016).

161 Younger Americans, we know, vote less . . . : File, Thom (2017).

161 Younger Americans, critics assert, also consume . . .": For more detailed data on news consumption by age, see Pew Research Center, *In Changing News Landscape* (2012).

161 And it is true: millennials are much less likely . . . : (Galston and Hendrickson, 2016). Party identification with the two major political parties, Democrats and Republicans, has declined in recent years. For example, 45 percent of young people age 18–29

identified as a Democrat in 2008, compared to only 37 percent in 2016. Whereas 26 percent of young people identified as Republican in 2008, that figure was 27 percent in 2016. By comparison, the percentage of young people who identify as Independent increased from 29 percent in 2008 to 35 percent in 2016.

161   Some studies, for example, suggest that . . . : Mitchell et al. (2016). A study by the Pew Research Center finds that young adults trust the information from national news media less than older adults and are also less likely to say that the national media do a good job of keeping them informed.

161   Others argue that apathy . . . : Kaiser (2014).

161   Additionally, millennials are frequently accused . . . : Stein (2013). For a critique of the view that millennials exhibit high levels of narcissism, see, for example, Brooke Lea Foster (2014).

161   Consequently, one scholar claims: Schement (2006)

162   Critics and social scientists alike: Turkle (2011).

162   Nan Lin, a longtime scholar in . . . : Lin, *Social Capital* (2001).

163   In a national survey of young adults . . . : For more on the study of how millennials' use of social media is evolving, see Rideout and Watkins (2018).

164   The shifting media landscape in which we . . . : Jenkins, Green, and Ford (2013).

164   Critics charge, for example, that social media . . . : Gladwell (2010).

164   This new repertoire was never more evident than . . . : Kang (2015).

166   By contrast, none of the major cable news channels . . . : Hitlin and Vogt (2014); Carr (2014).

166   Two researchers from Northeastern University . . . : Jackson and Welles (2015).

167   One of the first tweets related to . . . : Roy (2014).

167   "This story was put on the map . . .": Carr (2014).

167   This trend—citizens producing and circulating news . . . : For more on the agenda-setting function of the news media, see the origins of the theory in McCombs and Shaw (1972).

168   "We aren't born woke . . .": Mckesson, "Building Tools" (2016).

168   "It was 1 a.m. on August 16, 2014 . . .": Mckesson, "In Conversation" (2015).

168   "I packed a small bag . . .": Mckesson, "In Conversation" (2015).

169   "There is no one way to do this work . . .": Mckesson, "Building Tools" (2016).

169   "Twitter," Mckesson would later say, "was how I . . .": Monllos (2016).

169   "I remember when Trayvon Martin . . .": Mock (2016).

170   "I never criticize people who . . .": Mckesson, "Building Tools" (2016).

170   "I think that we'll continue to see . . .": Mckesson, "Building Tools" (2016).

171   "I replied to the tweet . . .": Tate (2017).

172   "People in positions of power . . .": Jazelle Hunt (2015).

173   "have been here all along . . .": Sinyangwe (2015).

173   "We could tell the story in another . . .": Sinyangwe (2015).

173   "One look at that map, in two seconds . . .": Captain (2017).

173   In 2015 fifty-nine of the sixty police departments . . . : Sinyangwe, Mckesson, and Packnett (2015).

176   "There are not enough full-time activists . . .": (Saxena 2017).

176   The contributions of Aditi Juneja . . . : The descriptions of Juenja's involvement with the Resistance Manual are based on author interviews conducted in the summer of 2017.

177   "What it [Twitter] hasn't done effectively . . .": Keefe (2017).

177   "[It] promotes crowd sourcing mentality . . .": Keefe (2017).

178   "When you think about Wikipedia . . .": Saxena (2017).

178   "network extensity" . . . : Lin, "Social Network" (1999).

Conclusion: Detroit Hustles Harder

185   "statistical significance" . . . : Aguilar and MacDonald (2015).

185   According to William Frey, a noted demographer . . . : Frey (2015).

185   "You can make a bigger impact . . .": Chaey (2013).

185   "Brooklynization of Detroit" . . . : Flanagin (2015).

186   Similar to venture funding in other locales . . . : Michigan Venture Capital Association (2017).

186   In its *2017 Scoring Tech Talent in North America* report . . . : CBRE (2017).

187   Once not even an afterthought . . . : For more on Detroit as a "brain gain," see CBRE, 2017. For another perspective on Detroit's "brain gain," see Florida, "Brain Gain in the Rustbelt" (2015).

187   Detroit, however, continues to be the poorest . . . : US Census Bureau, "Detroit City, Michigan" (2017).

187   The distribution of positive demographic . . . : Reese et al. (2017).

187   "stark contrast to what is happening . . .": Reese et al. (2017, 375).

187   The researchers at the Kirwan Institute . . . : For more details on the Kirwan Institute opportunity mapping analysis of Detroit, see Reece (2014).

188   "We want to bring people back . . .": Maynard (2014).

189   the need to create opportunities for entrepreneurs . . .: Ponyride (2016).

189   In this case, they were building . . .: Christakis and Fowler (2009).

190   "a study to see how the foreclosure . . .": Ponyride, "About Us: Our Focus" (2018).

192   design something to fill a need . . . : Veronika Scott, "Veronika" (2012).

192   "looked like a body bag": Veronika Scott, "Veronika" (2012).

192   "I spent eighty hours to sew trash bags . . .": Veronika Scott, "Veronika" (2012).

193   "We don't need coats . . .": Veronika Scott, "Veronika" (2012).

193   McInnis told me when . . . : Some of my description of the Empowerment Plan is based on an interview I conducted with Hailey McInnis in October 2017.

194   Rather than pass that cost on to the tenants . . . : Hooper (2018).

196   As Alexandria Ocasio-Cortez suggests . . . : Gonzalez-Ramirez (2018).

BIBLIOGRAPHY

Acemoglu, Daron, and David Autor. "Skills, Tasks and Technologies: Implications for Employment and Earnings." *Handbook of Labor Economics* 4 (2011): 1043–171.

Aguilar, Louis, and Christine MacDonald. "Detroit's White Population Up After Decades of Decline." *Detroit News*, September 17, 2015. https://www.detroitnews.com /story/news/local/detroit-city/2015/09/17/detroit-white-population-rises-census -shows/72371118.

Alexander, Karl L., Doris R. Entwisle, and Linda S. Olson. "Schools, Achievement, and Inequality: A Seasonal Perspective." *Educational Evaluation and Policy Analysis* 23, no. 2 (Summer 2001): 171–91.

Alexander, Leigh. "170,000 Devs Using Unity at Fifth Anniversary." *Gamasutra*, June 7, 2010. https://www.gamasutra.com/view/news/119793/170000_Devs_Using_Unity _At_Fifth_Anniversary.php.

Amis, Kelly, Kalimah Priforce, Denmark West, and Derrell Bradford. "*Code Oakland* Premiere in NYC: Panel Discussion." Moderated by Dennis Paul. Filmed in New York, NY, March 11, 2015. Video, 57:49. https://www.youtube.com/watch?v =UxUqzZocpEY.

Anderson, Chris. *The Long Tail: Why the Future of Business Is Selling Less of More*. New York: Hyperion, 2006.

Ashcraft, Catherine, Brad McLain, and Elizabeth Eger. *Women in Tech: The Facts*. National Center for Women and Information Technology, 2016.

Ault, Susanne. "Survey: YouTube Stars More Popular Than Mainstream Celebs Among U.S. Teens." *Variety*, August 5, 2014. https://variety.com/2014/digital /news/survey-youtube-stars-more-popular-than-mainstream-celebs-among -u-s-teens-1201275245.

Autor, David, Lawrence F. Katz, and Alan B. Krueger. "Computing Inequality: Have Computers Changed the Labor Market?" NBER Working Paper No. 5956. National Bureau of Economic Research, March 1997.

Barnes, Tom. "Meet the Rapper Who Recorded an Entire Album in the Apple Store." *Mic*, July 10, 2015. https://mic.com/articles/122061/meet-the-rapper-that-secretly -recorded-an-entire-album-in-the-apple-store#.KhqcFUZMr.

Barone, Matt. *"Dear White People* Director Justin Simien Wants to Change the Way We Talk about Movies." *Complex*, October 20, 2014. https://www.complex.com/pop -culture/2014/10/interview-dear-white-people-justin-simien.

Beaumont-Thomas, Ben. "No Studio? No Problem. Meet Prince Harvey, the Man Who Secretly Recorded an Album at the Apple Store." *Guardian*, July 12, 2015. https://www.theguardian.com/global/shortcuts/2015/jul/12/prince-harvey-rapper -secret-apple-album-apple-store-garageband.

Beer, Jeff. "Means of Production Subverts Advertising to Sell Americans on Socialism." *Fast Company*, July 26, 2018.

Bell, Daniel. *The Coming of Post-Industrial Society: A Venture in Social Forecasting*. New York: Basic Books, 1973.

Berkowitz, Joe. "The Awkward Ascent of *The Misadventures of Awkward Black Girl*." *Fast Company*, September 20, 2012.

Bernstein, Paula. *"Dear White People*: From Indiewire Project of the Year to Sundance Film Festival." *IndieWire*, December 5, 2013. https://www.indiewire.com/2013/12 /dear-white-people-from-indiewire-project-of-the-year-to-sundance-film-festival -32267.

Billboard Staff. "Billboard 200 Makeover: Album Chart to Incorporate Streams & Track Sales." *Billboard*, November 19, 2014. https://www.billboard.com/articles/columns /chart-beat/6320099/billboard-200-makeover-streams-digital-tracks.

Black Girls Code. "About Our Founder." 2018. http://www.blackgirlscode.com/about -bgc.html.

Bogost, Ian. *How to Do Things with Video Games*. Minneapolis: University of Minnesota Press, 2011.

Bordine, Kate. "Ponyride . . . a Home for Social Entrepreneurs Sharing Knowledge, Resources and Networks in Detroit's Corktown." Interview by Mark Maynard. March 13, 2014. http://markmaynard.com/2014/03/ponyride-a-home-for-social -entrepreneurs-sharing-knowledge-resources-and-networks-in-corktown.

Bosworth, Andrew. "What's the History of the Awesome Button (That Eventually Be-came the Like Button) on Facebook?" Quora, October 16, 2014. https://www .quora.com/Whats-the-history-of-the-Awesome-Button-that-eventually-became -the-Like-button-on-Facebook.

Bowles, Nellie. "A Dark Consensus about Screens and Kids Begins to Emerge in Silicon Valley." *New York Times*, October 26, 2018. https://www.nytimes.com/2018/10/26 /style/phones-children-silicon-valley.html.

———. "The Digital Gap between Rich and Poor Kids Is Not What We Expected." *New York Times*, October 26, 2018. https://www.nytimes.com/2018/10/26/style/digital -divide-screens-schools.html?action=click&module=RelatedLinks&pgtype=Article.

———. "Silicon Valley Nannies Are Phone Police for Kids. *New York Times*, October 26, 2018. https://www.nytimes.com/2018/10/26/style/silicon-valley-nannies.html?action =click&module=RelatedLinks&pgtype=Article.

Bretschneider, Ulrich, Katharina Knaub, and Enrico Wieck. "Motivations for Crowdfunding: What Drives the Crowd to Invest in Start-Ups?" In ECIS 2014

Proceedings, Tel Aviv, Israel. https://aisel.aisnet.org/ecis2014/proceedings /track05/6.

Brown, Tim. *Change by Design: How Design Thinking Transforms Organizations and Inspires Innovation.* New York: HarperBusiness, 2009.

Brynjolfsson, Erik, and Andrew McAfee. *The Second Machine Age: Work, Progress, and Prosperity in a Time of Brilliant Technologies.* New York: W. W. Norton, 2014.

Bump, Philip. "Tuesday Night Saw the Defining Upset of 2018." *Washington Post,* June 27, 2018. https://www.washingtonpost.com/news/politics/wp/2018/06/27 /tuesday-night-saw-the-defining-upset-of-2018/?utm_term=.4911cc932be2.

Caplan, Bryan. *The Case Against Education: Why the Education System Is a Waste of Time and Money.* Princeton, NJ: Princeton University Press, 2018.

———. "The World Might Be Better Off Without College for Everyone." *Atlantic,* January/February, 2018.

Captain, Sean. "How Trump's Opponents Are Crowdsourcing the Resistance." *Fast Company,* January 31, 2017. https://www.fastcompany.com/3067643/how-trumps -opponents-are-crowdsourcing-the-resistance.

Caramanica, Jon. "The Rowdy World of Rap's New Underground." *New York Times,* June 22, 2017. https://www.nytimes.com/2017/06/22/arts/music/soundcloud-rap -lil-pump-smokepurrp-xxxtentacion.html.

———. "Two SoundCloud Rap Outlaws Push Boundaries from the Fringes." *New York Times,* March 21, 2018. https://www.nytimes.com/2018/03/21/arts/music /xxxtentacion-question-6ix9ine-day69-review.html.

Carless, Simon. "OTEE Releases Unity 1.1. Game Engine." *Gamasutra,* August 23, 2005. https://www.gamasutra.com/view/news/97262/OTEE_Releases_Unity_11 _Game_Engine.php.

Carr, David. "View of #Ferguson Thrust Michael Brown Shooting to National Atten- tion." *New York Times,* August 17, 2014. https://www.nytimes.com/2014/08/18 /business/media/view-of-ferguson-thrust-michael-brown-shooting-to-national -attention.html.

Caulfield, Keith. "Travis Scott's 'Astroworld' Returns to No. 1 on Billboard 200 Chart, 6ix9ine's 'Dummy Boy' Debuts at No. 2." *Billboard,* December 2, 2018. https:// www.billboard.com/articles/columns/chart-beat/8487688/travis-scott-astroworld -returns-to-no-1-billboard-200.

CBRE Research. *Scoring Tech Talent in North America 2017.* https://www.bestchamber .com/uploads/5/5/2/2/55223453/2017_scoring_tech_talent.pdf.

Chaey, Christina. "Detroit Is Going Bankrupt—but Its Tech Community Is Going Strong." *Fast Company,* July 18, 2013. https://www.fastcompany.com/3014543 /detroit-is-going-bankrupt-but-its-tech-community-is-going-strong.

Chan, Kathy H. "I Like This." Facebook, February 9, 2009. https://www.facebook .com/notes/facebook/i-like-this/53024537130.

Chang, Richard. "Half of High School Seniors Lack Access to Computer Science." *Journal,* August 17, 2016. https://thejournal.com/articles/2016/08/17/half-of-high -school-seniors-lack-access-to-computer-science.aspx.

Charles, Simon. "OverTheEdge Announces Unity 1.0.2 Release." *Gamasutra,* July 14, 2005. http://www.gamasutra.com/view/news/5937/OverTheEdge_Announces _Unity_102_Release.php.

Chávez, Aida, and Ryan Grim. "A Primary Against the Machine: A Bronx Activist Looks to Dethrone Joseph Crowley, the King of Queens." *Intercept*, May 22, 2018. https://theintercept.com/2018/05/22/joseph-crowley-alexandra-ocasio-cortez-new-york-primary.

Chesbrough, Henry. "The Era of Open Innovation." *MIT Sloan Management Review* 44, no. 3 (Spring 2003): 35–41.

Chetty, Raj, David Grusky, Maximilian Hell, Nathaniel Hendren, Robert Manduca, and Jimmy Narang. "The Fading American Dream: Trends in Absolute Income Mobility Since 1940." The Equality of Opportunity Project, Stanford University, 2016. http://www.equality-of-opportunity.org/assets/documents/abs_mobility_summary.pdf.

Christakis, Nicholas A., and James H. Fowler. *Connected: The Surprising Power of Our Social Networks and How They Shape Our Lives*. New York: Little, Brown, 2009.

Condon, Meghan, and Matthew Holleque. "Entering Politics: General Self-Efficacy and Voting Behavior among Young People." *Political Psychology* 34, no. 2 (April 2013): 167–81.

Corkery, Michael. "Toys 'R' Us Says It Will Close or Sell All U.S. Stores." *New York Times*, March 14, 2018. https://www.nytimes.com/2018/03/14/business/toy-r-us-closing.html.

Coscarelli, Joe. "Lil Uzi Vert Debuts at No. 1, Leading a SoundCloud Wave." *New York Times*, September 4, 2017. https://www.nytimes.com/2017/09/04/arts/music/lil-uzi-vert-xxxtentacion-billboard-chart.html.

Crawford, Matthew B. *Shop Class as Soulcraft: An Inquiry into the Value of Work*. New York: Penguin, 2009.

Cueva, Olivia. "Bay Area Hackathon Brings Tech to Youth of Color." KALW, San Francisco, March 23, 2015. http://www.kalw.org/post/bay-area-hackathon-brings-tech-youth-color#stream/0.

Currid, Elizabeth. *The Warhol Economy: How Fashion, Art, and Music Drive New York City*. Princeton, NJ: Princeton University Press, 2008.

Daalder, Marc. "How Detroit Filmmakers Helped a Socialist Unseat a New York Lawmaker." *Detroit Free Press*, June 28, 2018. https://www.freep.com/story/news/politics/2018/06/28/alexandria-ocasio-cortez-joe-crowley/742202002.

Dasgupta, Nilanjana. " 'Belonging' Can Help Keep Talented Female Students in STEM Classes." Interview by the National Science Foundation. August 26, 2016. https://www.nsf.gov/discoveries/disc_summ.jsp?cntn_id=189603.

DeRuy, Emily. "Why This Man Happily Turns Students into 'Hackers.' " *Atlantic*, July 15, 2015. https://www.theatlantic.com/politics/archive/2015/07/why-this-man-happily-turns-students-into-hackers/432321.

Directors Guild of America. "DGA 2016–17 Episodic TV Director Diversity Report." November 14, 2017. https://www.dga.org/News/PressReleases/2017/171114-Episodic-Television-Director-Diversity-Report.aspx.

Duffy, Brook Erin. *(Not) Getting Paid To Do What You Love: Gender, Social Media, and Aspirational Work*. New Haven, CT: Yale University Press.

Dweck, Carol S. *Mindset: The New Psychology of Success*. New York: Random House, 2006.

Eler, Alicia. "Issa Rae's Long Road from YouTube to HBO." *Daily Dot*, August 28, 2016. https://www.dailydot.com/upstream/issa-rae-insecure-hbo.

Endler, Michael. "Microsoft Workforce Grows More Diverse." *Information Week*, October 6, 2014. https://www.informationweek.com/strategic-cio/team-building-and-staffing/microsoft-workforce-grows-more-diverse/d/d-id/1316401.

Entertainment Software Association. "Women Comprise Nearly Half of Gamer Population." June 11, 2013. http://www.theesa.com/article/women-comprise-nearly-half-gamer-population.

Ericson, Barbara. "Detailed Race and Gender Information 2017." January 8, 2018. http://home.cc.gatech.edu/ice-gt/599.

File, Thom. "Voting in America: A Look at the 2016 Presidential Election." *Census Blogs*. US Census Bureau, 2017. https://www.census.gov/newsroom/blogs/random-samplings/2017/05/voting_in_america.html.

Flanagin, Jake. "The Brooklynization of Detroit Is Going to Be Terrible for Detroiters." *Quartz*, July 15, 2015. https://qz.com/453531/the-brooklynization-of-detroit-is-going-to-be-terrible-for-detroiters.

Florida, Richard. "The Amazon Model: The Rise of Urban Start-Ups in Smaller Tech Hubs." *CityLab*, August 27, 2013. https://www.citylab.com/life/2013/08/amazon-model-rise-urban-start-ups-smaller-tech-hubs/6405.

———. "Brain Gain in the Rustbelt." *CityLab*, August 31, 2015. https://www.citylab.com/equity/2015/08/brain-gain-in-the-rustbelt/402922.

———. *The Rise of the Creative Class: And How It's Transforming Work, Leisure, Community and Everyday Life*. New York: Basic Books, 2002.

Florida, Richard, and Karen M. King. "Rise of the Urban Startup Neighborhood: Mapping Micro-Clusters of Venture Capital-Based Startups." Toronto: University of Toronto Martin Prosperity Institute, June 14, 2016. http://martinprosperity.org/content/rise-of-the-urban-startup-neighborhood.

Florida State University. "Under Challenge: Girls' Confidence Level, Not Math Ability Hinders Path to Science Degrees." Phys.org, April 6, 2017. https://phys.org/news/2017-04-girls-confidence-math-ability-hinders.html.

Foster, Brooke Lea. "The Persistent Myth of the Narcissistic Millennial." *Atlantic*, November 19, 2014.

Foster, Elizabeth. "GoldieBlox Builds Brand Beyond Construction Sets." *Kidscreen*, September 27, 2016. http://kidscreen.com/2016/09/27/goldieblox-builds-brand-beyond-construction-sets.

Frey, William H. "More Big Cities Are Gaining White Population, Census Data Show." *The Avenue*. Brookings Institution, October 1, 2015. https://www.brookings.edu/blog/the-avenue/2015/10/01/more-big-cities-are-gaining-white-population-census-data-show.

———. "The US Will Become 'Minority White' in 2045, Census Projects." *The Avenue*. Brookings Institution, March 14, 2018. https://www.brookings.edu/blog/the-avenue/2018/03/14/the-us-will-become-minority-white-in-2045-census-projects.

Friedlander, Joshua P. *News and Notes on 2017 RIAA Revenue Statistics*. Record Industry Association of America, 2018. https://www.riaa.com/wp-content/uploads/2018/03/RIAA-Year-End-2017-News-and-Notes.pdf.

Fry, Richard. "Millennials Aren't Job-Hopping Any Faster Than Generation X Did." *Fact Tank*, Pew Research Center, April 19, 2017. http://www.pewresearch.org/fact-tank/2017/04/19/millennials-arent-job-hopping-any-faster-than-generation-x-did.

————. "Millennials Are the Largest Generation in the U.S. Labor Force." *Fact Tank*. Pew Research Center, April 11, 2018. http://www.pewresearch.org/fact-tank/2018/04/11/millennials-largest-generation-us-labor-force.

Gabbatt, Adam. "*Dear White People* Director—We're Definitely Not a Post-Racial Society." *Guardian*, January 19, 2014. https://www.theguardian.com/film/2014/jan/19/dear-white-people-director-post-racial-society.

Galston, William A., and Clara Hendrickson. "How Millennials Voted This Election." *FixGov*. Brookings Institution, November 21, 2016. https://www.brookings.edu/blog/fixgov/2016/11/21/how-millennials-voted.

Gervais, Martin, Nir Jaimovich, Henry E. Siu, and Yaniv Yedid-Levi. "What Should I Be When I Grow Up? Occupations and Unemployment over the Life Cycle." NBER Working Paper No. 20628. National Bureau of Economic Research, October 2014. https://www.nber.org/papers/w20628.pdf.

Gibbs, Samuel. "Google Employs Just 30% Women and 2% Black People, Report Shows." *Guardian*, May 29, 2014. https://www.theguardian.com/technology/2014/may/29/google-diversity-women-black-people.

Gladwell, Malcolm. "Small Change: Why the Revolution Will Not Be Tweeted." Annals of Innovation. *New Yorker*, October 4, 2010.

Goldmacher, Shane. "An Upset in the Making: Why Joe Crowley Never Saw Defeat Coming." *New York Times*, June 27, 2018. https://www.nytimes.com/2018/06/27/nyregion/ocasio-cortez-crowley-primary-upset.html.

Goldmacher, Shane, and Jonathan Martin. "Alexandria Ocasio-Cortez Defeats Joseph Crowley in Major Democratic House Upset." *New York Times*, June 26, 2018. https://www.nytimes.com/2018/06/26/nyregion/joseph-crowley-ocasio-cortez-democratic-primary.html.

Gonzalez-Ramirez, Andrea. "Alexandria Ocasio-Cortez Just Defeated One of the Most Powerful House Democrats." June 26, 2018. Refinery29. https://www.refinery29.com/2018/06/202947/alexandria-ocasio-cortez-wins-new-york-14th-district-primary.

Google. *Google Workforce Composition 2014–2018*. https://static.googleusercontent.com/media/diversity.google/en//static/pdf/Google_Workforce_Composition_2014-2018.pdf.

Google Inc. & Gallup Inc. *Trends in the State of Computer Science in U.S. K–12 Schools*. https://services.google.com/fh/files/misc/trends-in-the-state-of-computer-science-report.pdf.

Granovetter, Mark. *Getting a Job: A Study of Contacts and Careers*. 2nd ed. Chicago: University of Chicago Press, 1995.

————. "The Strength of Weak Ties." *American Journal of Sociology* 78, no. 6 (May 1973): 1360–80.

Greenwald, Anthony G., and Mahzarin R. Banaji. "Implicit Social Cognition: Attitudes, Self-Esteem, and Stereotypes." *Psychological Review* 102, no. 1 (January 1995): 4–27.

Hacker, Jacob S. *The Great Risk Shift: The Assault on American Jobs, Families, Health Care, and Retirement and How You Can Fight Back*. New York: Oxford University Press, 2006.

Hargittai, Eszter. "Second-Level Digital Divide: Differences in People's Online Skills." *First Monday* 7, no. 4 (April 2002). http://firstmonday.org/ojs/index.php/fm/article/view/942/864.

Hargittai, Eszter, and Amanda Hinnant. "Digital Inequality: Differences in Young Adults' Use of the Internet." *Communication Research* 35, no. 5 (October 2008): 602–21.

Harvey, David. *The Condition of Postmodernity: An Enquiry into the Origins of Social Change*. Malden, MA: Blackwell Publishing, 1990.

Harvey, Prince. "Recorded My Album PHATASS (Prince Harvey at the Apple Store: Soho) at an Apple Store." Facebook, February 22, 2016. https://www.facebook.com /princeharveylove/videos/i-recorded-my-album-at-an-apple-store-this-is-my-first -music-/1693077680946260.

Hayes, Dade, and Dawn C. Chmlielewski. "Netflix Shares Dive Amid Exec's Exit and Looming Diversity Questions." *Deadline Hollywood*, June 25, 2018. https://deadline .com/2018/06/netflix-shares-dive-after-exec-exit-diversity-questions-1202416808.

Heckman, James J. "Policies to Foster Human Capital," *Research in Economics* 54, no. 1 (March 2000): 3–56.

Heffernan, Virginia, and Tom Zeller. "*Lonely Girl* (and Friends) Just Wanted Movie Deal." *New York Times*, September 12, 2006. https://www.nytimes.com/2006/09/12 /technology/12cnd-lonely.html.

Hickey, Walt. "*Black Panther* Is Groundbreaking, but It's Shuri Who Could Change the World." *FiveThirtyEight*, February 18, 2018. https://fivethirtyeight.com/features /black-panther-is-groundbreaking-but-its-shuri-who-could-change-the-world.

Hill, Logan. "10 Directors to Watch: Justin Simien Puts *White People* in Perspective." *Variety*, December 17, 2013. https://variety.com/2013/film/features/10-directors-to -watch-justin-simien-puts-white-people-in-perspective-1200970719.

Hitlin, Paul, and Nancy Vogt. "Cable, Twitter Picked Up Ferguson Story at a Similar Clip." *Fact Tank*. Pew Research Center, August 20, 2014. http://www.pewresearch .org/fact-tank/2014/08/20/cable-twitter-picked-up-ferguson-story-at-a-similar-clip.

Hooper, Ryan Patrick. "Ponyride Plans to Sell Corktown Building and Move to Make Art Work Near New Center." *Detroit Free Press*, February 9, 2018. https://www .freep.com/story/entertainment/arts/2018/02/09/ponyride-corktown-move-detroit -make-art-work-recycle-here/321501002.

Hoover, Betsy. "How Does Cambridge Analytica Flap Compare With Obama's Campaign Tactics?" Interview by Lulu Garcia-Navarro. *Weekend Edition Sunday*, National Public Radio, March 25, 2018. https://www.npr.org/2018/03/25/596805347 /how-does-cambridge-analytica-flap-compare-with-obama-s-campaign-tactics.

Horgan, Maya. "Texas Hackers to Fight Ebola, Fire and Water Contamination with African SMS." *Huffington Post*, December 6, 2014. https://www.huffingtonpost.com /maya-horgan/texas-hackers-to-fight-eb_b_6069490.html.

Hunt, Darnell. *Race in the Writers' Room: How Hollywood Whitewashes the Stories That Shape America*. Color of Change, 2017. https://hollywood.colorofchange.org.

Hunt, Jazelle. "Ferguson—One Year Later." *Seattle Medium*, August 12, 2015. http:// seattlemedium.com/ferguson-one-year-later.

"Interview with Internet Sensation Issa Rae." Summary of interview by Amanda de Cadenet. *Hello Giggles*. https://hellogiggles.com/reviews-coverage/interview-with -internet-sensation-issa-rae.

Irani, Lilly. " 'Design Thinking': Defending Silicon Valley at the Apex of Global Labor Hierarchies." *Catalyst: Feminism, Theory, Technoscience* 4, no. 1 (2018): 1–19.

Israni, Ellora Thadaney. "When an Algorithm Helps Send You to Prison." *New York Times*, October 26, 2017. https://www.nytimes.com/2017/10/26/opinion/algorithm-compas-sentencing-bias.html.

Jackson, Sarah J., and Brooke Foucault Welles. "#Ferguson Is Everywhere: Initiators in Emerging Counterpublic Networks." *Information, Communication & Society* 19, no. 3 (2015): 397–418.

Jacobs, Jane. *The Death and Life of Great American Cities*. 50th anniversary ed. New York: Modern Library, 2011.

Jenkins, Henry, Joshua Green, and Sam Ford. *Spreadable Media: Creating Value and Meaning in a Networked Culture*. New York: New York University Press, 2013.

Jilani, Zaid. "How a Ragtag Group of Socialists Filmmakers Produced One of the Most Viral Campaign Ads of 2018." *Intercept*, June 5, 2018.https://theintercept.com/2018/06/05/ocasio-cortez-new-york-14th-district-democratic-primary-campaign-video.

Jilani, Zaid, and Ryan Grim. "Data Suggest That Gentrifying Neighborhoods Powered Alexandria Ocasio-Cortez's Victory." *Intercept*, July 1, 2018. https://theintercept.com/2018/07/01/ocasio-cortez-data-suggests-that-gentrifying-neighborhoods-powered-alexandria-ocasio-cortesz-victory-over-the-democratic-establishment.

Kaiser, Robert G. "The Bad News About the News." The Brookings Essay. Brookings Institution, October 16, 2014. http://csweb.brookings.edu/content/research/essays/2014/bad-news.html#.

Kang, Jay Caspian. "Our Demand Is Simple: Stop Killing Us." *New York Times Magazine*, May 4, 2015. https://www.nytimes.com/2015/05/10/magazine/our-demand-is-simple-stop-killing-us.html.

Kantor, Jodi, and David Streitfeld. "Inside Amazon: Wrestling Big Ideas in a Bruising Workplace." *New York Times*, August 15, 2015. https://www.nytimes.com/2015/08/16/technology/inside-amazon-wrestling-big-ideas-in-a-bruising-workplace.html.

Karim, Jawed. "Jawed Karim, Illinois Commencement 2007, pt2." YouTube. Posted June 5, 2007. Video: 7:18. https://www.youtube.com/watch?v=24yglUYbKXE.

Katz, Bruce, and Julie Wagner. *The Rise of Innovation Districts: A New Geography of Innovation in America*. Metropolitan Policy Program at Brookings. Brookings Institution, May 2014.

Katz, Lawrence B., and Alan B. Krueger. *The Rise and Nature of Alternative Work Arrangements in the United States, 1995–2015*. NBER Working Paper No. 22667. National Bureau of Economic Research, September 2016.

Kaushal, Neeraj, Katherine Magnuson, and Jane Waldfogel. "How Is Family Income Related to Investments in Children's Learning?" In *Whither Opportunity? Rising Inequality, Schools, and Children's Life Chances*, edited by Greg J. Duncan and Richard J. Murnane, 187–206. New York: Russell Sage Foundation, 2011.

Kavilanz, Parija. "GoldieBlox Ad Makes Super Bowl History," CNN Business, January 31, 2014. https://money.cnn.com/2014/01/30/smallbusiness/super-bowl-ad-intuit-goldieblox/index.html.

Keefe, Josh. "How Black Lives Matter Will Protest Trump: Activists Release 'Resistance Manual' for GOP Administration." *International Business Times*, January 18, 2017. https://www.ibtimes.com/how-black-lives-matter-will-protest-trump-activists-release-resistance-manual-gop-2477186.

Kelly, Kevin. "Better Than Human: Why Robots Will—and Must—Take Our Jobs." *Wired*, December 24, 2012.

Kesar, Shalini. *Closing the STEM Gap: Why STEM Classes and Careers Still Lack Girls and What We Can Do About It.* Microsoft, 2018. https://query.prod.cms.rt.microsoft .com/cms/api/am/binary/RE1UMWz.

Keyani, Pedram. "Stay Focused and Keep Hacking." Facebook, May 23, 2012. https:// www.facebook.com/notes/facebook-engineering/stay-focused-and-keep-hacking /10150842676418920.

Kimble, Julian. "The Viral, Meme-Inspiring #InsecureHBO Hashtag Is as Much a Smash Hit as 'Insecure' Itself." *Undefeated*, September 3, 2017. https://the undefeated.com/features/insecure-hbo-viral-hashtag.

Knopper, Steve. "A Hip-Hop Signing Frenzy Sends New Record Deal Prices Soaring." *Billboard*, March 29, 2018. https://www.billboard.com/articles/business/8272682 /hip-hop-signing-frenzy-record-deal-prices-soaring.

Kuehn, Kathleen, and Thomas F. Corrigan. "Hope Labor: The Role of Employment Prospects in Online Social Production." *Political Economy of Communication* 1, no. 1 (2013): 9–25.

Kuppuswamy, Venkat, and Barry L. Bayus. "Crowdfunding Creative Ideas: The Dynamics of Projects Backers in Kickstarter," SSRN Working Paper. March 17, 2013. https://papers.ssrn.com/sol3/papers.cfm?abstract_id=2234765.

Langdon, David, George McKittrick, David Beede, Beethika Khan, and Mark Doms. *STEM: Good Jobs Now and for the Future.* Washington, DC: Department of Commerce, July 2011.

Leckart, Steven. "The Hackathon Fast Track, from Campus to Silicon Valley." *New York Times*, April 6, 2015. https://www.nytimes.com/2015/04/12/education/edlife/the -hackathon-fast-track-from-campus-to-silicon-valley.html.

Levy, Frank, and Richard J. Murnane. *The New Division of Labor: How Computers Are Creating the Next Job Market.* Princeton, NJ: Princeton University Press, 2004.

Liming, Drew, and Dennis Vilorio. "Work for Play: Careers in Video Game Development." *Occupational Outlook Quarterly* 55, no. 3 (Fall 2011): 2–11.

Lin, Nan. *Social Capital: A Theory of Social Structure and Action.* Cambridge, UK: Cambridge University Press, 2001.

———. "Social Network and Status Attainment." *Annual Review of Sociology* 25, no. 1 (August 1999): 467–87.

Ljung, Alex. "Two Laptops and a Mobile Phone: Startup Music from SoundCloud's Alex Ljung." Interview by Noah Robischon. *Fast Company*, October 19, 2011. Video, 4:14. https://www.fastcompany.com/1788740/two-laptops-and-mobile-phone-startup -music-soundclouds-alex-ljung.

Lott, Eric. *Love and Theft: Blackface Minstrelsy and the American Working Class.* New York: Oxford University Press, 1993.

Maheshwari, Sapna. "On YouTube Kids, Startling Videos Slip Past Filters." *New York Times*, November 4, 2017. https://www.nytimes.com/2017/11/04/business/media /youtube-kids-paw-patrol.html.

———. "YouTube Is Improperly Collecting Children's Data, Consumer Groups Say." *New York Times*, April 9, 2018. https://www.nytimes.com/2018/04/09/business /media/youtube-kids-ftc-complaint.html.

Manyika, James, Susan Lund, Michael Chui, Jacques Bughin, Jonathan Woetzel, Parul Batra, Ryan Ko, and Saurabh Sanghvi. *Jobs Lost, Jobs Gained: Workforce Transitions in a Time of Automation*. McKinsey & Company, McKinsey Global Institute, December 2017.

Marker, Scott. "How Many Jobs Will the Average Person Have in His or Her Life-time?" LinkedIn, February 22, 2015. https://www.linkedin.com/pulse/how-many -jobs-average-person-have-his-her-lifetime-scott-marker.

McAlone, Nathan. "These Are the 18 Most Popular YouTube Stars in the World—and Some Are Making Millions." *Business Insider*, March 7, 2017. https://www.business insider.com/most-popular-youtuber-stars-salaries-2017/#no-16-jacksepticeye -148-million-subscribers-3.

McBride, Stormy. "Global Coworking Forecast: 30,432 Spaces and 5.1 Million Members by 2022." Global Coworking Unconference Conference, December 18, 2017. https:// gcuc.co/2018-global-coworking-forecast-30432-spaces-5-1-million-members-2022.

McCombs, Maxwell E., and Donald L. Shaw. "The Agenda-Setting Function of Mass Media." *Public Opinion Quarterly* 36, no. 2 (Summer 1972): 176–87.

Mckesson, DeRay. "Building Tools for Digital Activism." Interview by Kwame Opam. *Verge*, November 29, 2016. https://www.theverge.com/a/verge-2021/deray -mckesson-interview-black-lives-matter-digital-activism.

———. "In Conversation with DeRay Mckesson." Interview by Rembert Browne. *Intelligencer*, November 22, 2015. http://nymag.com/intelligencer/2015/11/conversation -with-deray-mckesson.html.

Michigan Venture Capital Association. *2017 Detroit Entrepreneurial Study: How Detroit Is Quickly Becoming a Center of Entrepreneurial and Investment Activity in Michigan*. http://michiganvca.org/wp-content/uploads/2017/06/2017-MVCA-Detroit -Entrepreneurial-Study2.pdf.

"Millennials Significantly Outpacing Other Age Groups for Taking On Side Gigs," CareerBuilder, September 29, 2016. https://www.careerbuilder.com/share/aboutus /pressreleasesdetail.aspx?ed=12%2F31%2F2016&id=pr968&sd=9%2F29%2F2016.

Miller, Claire Cain. "Google Releases Employee Data, Illustrating Tech's Diversity Challenge." *New York Times*, May 28, 2014. https://bits.blogs.nytimes.com/2014/05 /28/google-releases-employee-data-illustrating-techs-diversity-challenge.

Misulonas, Joseph. "Press Start to Begin?" *Texas Monthly*, March 6, 2014. https://www .texasmonthly.com/articles/press-start-to-begin.

Mitchell, Amy. "Younger Adults More Likely Than Their Elders to Prefer Reading News." *Fact Tank*. Pew Research Center, October 6, 2016. http://www.pewresearch .org/fact-tank/2016/10/06/younger-adults-more-likely-than-their-elders-to-prefer -reading-news.

Mitchell, Amy, Jeffrey Gottfried, Michael Barthel, and Elisa Shearer. *The Modern News Consumer: News Attitudes and Practices in the Digital Era*. Washington, DC: Pew Research Center, 2016.

Mock, Janet. "Vested Interests: Why DeRay Mckesson Matters." *Advocate*, February 25, 2016. https://www.advocate.com/current-issue/2016/2/25/janet-mock-why-deray -mckesson-matters.

Mohan, Neal, and Robert Kyncl. "Additional Changes to the YouTube Partner Program (YPP) to Better Protect Creators." *YouTube Creator Blog*, January 16, 2018.

Monllos, Kristina. "How DeRay Mckesson Turned Social Media into a Powerful Tool for Social Justice." *Adweek*, October 31, 2016, 20–24.

Moretti, Enrico. *The New Geography of Jobs*. New York: Houghton Mifflin Harcourt, 2012.

MSNBC Originals. "Swimming in Their Genius: How #YesWeCode Teaches Tech." December 3, 2014. Video, 9:35. http://www.msnbc.com/msnbc/watch/swimming-in-their-genius-366728259749?v=raila&.

Musu-Gillette, Lauren, Cristobal de Brey, Joel McFarland, William Hussar, William Sonnenberg, and Sidney Wilkinson-Flicker. "Graduation Rates from First Institution Attended for First-Time, Full-Time Bachelor's Degree-Seeking Students at 4-Year Postsecondary Institutions, by Race/Ethnicity and Time to Completion: Starting Cohort Year 2008." In *Status and Trends in the Education of Racial and Ethnic Groups*, figure 21.1. Washington, DC: US Department of Education, National Center for Education Statistics, July 2017. https://nces.ed.gov/programs/raceindicators/indicator_red.asp#1.

———. "Indicator 6: Elementary and Secondary Enrollment." In *Status and Trends in the Education of Racial and Ethnic Groups*. Washington, DC: US Department of Education, National Center for Education Statistics, July 2017. https://nces.ed.gov/programs/raceindicators/indicator_rbb.asp.

———. "Indicator 10: Mathematics Achievement." In *Status and Trends in the Education of Racial and Ethnic Groups*. Washington, DC: US Department of Education, National Center for Education Statistics, July 2017. https://nces.ed.gov/programs/raceindicators/indicator_rcb.asp.

———. "Indicator 18: College Participation Rates." In *Status and Trends in the Education of Racial and Ethnic Groups*. Washington, DC: US Department of Education, National Center for Education Statistics, July 2017. https://nces.ed.gov/programs/raceindicators/indicator_rea.asp.

———. "Indicator 21: Postsecondary Graduation Rates." In *Status and Trends in the Education of Racial and Ethnic Groups*. Washington, DC: US Department of Education, National Center for Education Statistics, July 2017. https://nces.ed.gov/programs/raceindicators/indicator_red.asp#1.

———. "Indicator 24: STEM Degrees." In *Status and Trends in the Education of Racial and Ethnic Groups*. Washington, DC: US Department of Education, National Center for Education Statistics, July 2017. https://nces.ed.gov/programs/raceindicators/indicator_reg.asp.

———. "Total College Enrollment Rates of 18- to 24-Year-Olds in Degree-Granting Institutions, by Race/Ethnicity: 2015." In *Status and Trends in the Education of Racial and Ethnic Groups*, figure 18.2. Washington, DC: US Department of Education, National Center for Education Statistics, July 2017. https://nces.ed.gov/programs/raceindicators/indicator_rea.asp#2.

Naasel, Kenrya Rankin. "How Issa Rae Went from Awkward Black Girl to Indie TV Producer." *Fast Company*, September 25, 2014. https://www.fastcompany.com/3036176/how-issa-rae-went-from-awkward-black-girl-to-indie-tv-producer.

Narvin, Matthew. "He Made a Secret Album in an Apple Store." *Daily Beast*, July 5, 2015. https://www.thedailybeast.com/he-made-a-secret-album-in-an-apple-store.

National Center for Women and Information Technology. "NCWIT Fact Sheet." National Science Foundation, 2017. https://www.ncwit.org/ncwit-fact-sheet.

National Telecommunications and Information Administration. *Falling Through the Net: Defining the Digital Divide*. Washington, DC: US Department of Commerce, 1999.

———. *Falling Through the Net: A Survey of the "Haves" and "Have Nots" in Rural and Urban America*. Washington, DC: US Department of Commerce, 1995.

———. *Falling Through the Net II: New Data on the Digital Divide*. A Report on the Telecommunications and Information Technology Gap in America. Washington, DC: US Department of Commerce, 1998.

Neff, Gina. *Venture Labor: Work and the Burden of Risk in Innovative Industries*. Cambridge, MA: MIT Press, 2012.

Newman, Andy, Vivian Wang, and Luis Ferré-Sadurní. "Alexandria Ocasio-Cortez Emerges as a Political Star." *New York Times*, June 27, 2018. https://www.nytimes.com/2018/06/27/nyregion/alexandria-ocasio-cortez-bio-profile.html.

Nicolaou, Anna. "SoundCloud on Track for Growth After Financial Rescue." *Financial Times*, April, 1, 2018. https://www.ft.com/content/2985ef90-3561-11e8-8eee-e06bdeo1c544.

Nosnitsky, Andrew. "Odd Future: The Billboard Cover Story." *Billboard*, March 11, 2011.

Obama, Barack. "Remarks by President Obama in Town Hall with Young Leaders of the UK." London, April 23, 2016. https://obamawhitehouse.archives.gov/the-press-office/2016/04/23/remarks-president-obama-town-hall-young-leaders-uk.

Ocasio-Cortez, Alexandria. "Alexandria Ocasio-Cortez Tackles New York Inequality." Interview by Emma Vigeland. *The Young Turks*, April 3, 2018. https://www.youtube.com/watch?v=cZ_qOQlQCio.

———. "Political Newcomer Alexandria Ocasio-Cortez on Her Upset and the Road Ahead." Interview by Joe Scarborough and Mika Brzezinski. *Morning Joe*. MSNBC, June 27, 2018. Video: 15:11. https://www.youtube.com/watch?v=AUb-QB8twcA.

Pacione, Chris. "Evolution of the Mind: A Case for Design Literacy." *Interactions* 17, no. 2 (March/April 2010): 6–11.

Parkin, Simon. "Don't Stop: The Game That Conquered Smartphones." *New Yorker*, June 7, 2013.

Pearce, Katy E., and Ronald E. Rice. "Digital Divides from Access to Activities: Comparing Mobile and Personal Computer Internet Users." *Journal of Communication* 63, no. 4 (August 2013): 721–44.

Pearl, Diana. "This 24-Year-Old Was a Patron at Alexandria Ocasio-Cortez's Bar. Now, She's Her Campaign Designer." *Adweek*, August 6, 2018. https://www.adweek.com/brand-marketing/a-24-year-old-is-the-woman-behind-alexandria-ocasio-cortezs-campaign-design.

Peoples, Lindsay. "Issa Rae On Making Black Experiences 'Regular' Events on TV." *The Cut*, October 4, 2016. https://www.thecut.com/2016/10/issa-rae-on-making-misadventures-of-awkward-black-girl.html.

Perez-Felkner, Lara, Samantha Nix, and Kirby Thomas. "Gendered Pathways: How Mathematics Ability Beliefs Shape Secondary and Postsecondary Course and Degree Field Choices. *Frontiers in Psychology* 8, no. 386 (April 6, 2017). doi:10.3389/fpsyg.2017.00386.

Pew Research Center. "About 6 in 10 Young Adults in U.S. Primarily Use Online Streaming to Watch TV." *Fact Tank*. Pew Research Center, September 13, 2017.

http://www.pewresearch.org/fact-tank/2017/09/13/about-6-in-10-young-adults
-in-u-s-primarily-use-online-streaming-to-watch-tv.

———. *In Changing News Landscape, Even Television Is Vulnerable*. September 27, 2012.
http://www.people-press.org/2012/09/27/in-changing-news-landscape-even
-television-is-vulnerable.

Pitterson, Leslie. "Where's the Black Version of Liz Lemon?" *TheGrio*, April 19, 2010.
https://thegrio.com/2010/04/19/wheres-the-black-version-of-liz-lemon.

Ponyride. "About Us: Our Focus." 2018. https://www.ponyride.org/start-your-own.

Primack, Brian A., et al. "Social Media Use and Perceived Social Isolation Among
Young Adults in the U.S." *American Journal of Preventive Medicine* 53, no. 1 (July
2017): 1–8. https://doi.org/10.1016/j.amepre.2017.01.010.

Putnam, Robert D. *Bowling Alone: The Collapse and Revival of American Community*. New
York: Simon & Schuster, 2000.

Quinn, Eithne. *A Piece of the Action: Race and Labor in Post–Civil Rights American Cinema*.
New York: Columbia University Press, forthcoming.

Rae, Issa. "Awkward Black Girl: A Conversation with Issa Rae." Austin Film Festival,
June 4, 2016. Video, 26:46. https://www.youtube.com/watch?v=3Yqo3BjK2XU.

———. "HAATBP Keynote: Issa Rae." Streamed live on September 24, 2014, from the
One Club for Creativity in New York, NY. Video, 49:54. https://www.youtube.com
/watch?v=gAqboLHK5Lw.

———. "Issa Rae, Creator of *Awkward Black Girl*, Felt Like Her Voice Was Missing
from Pop Culture—So Here's What She Did." Interview by Emma Gray. *Huffington Post*, November 5, 2013. https://www.huffingtonpost.com/2013/11/05/issa-rae
-awkward-black-girl_n_4209313.html.

———. "Issa Rae of *Awkward Black Girl* on the Future of the Web Series." Interview by
Lily Rothman. *Time*, July 10, 2012. http://entertainment.time.com/2012/07/10
/issa-rae-of-awkward-black-girl-on-the-future-of-the-web-series.

———. *The Misadventures of Awkward Black Girl*. New York: Atria, 2015.

———. *The Misadventures of Awkward Black Girl*. Kickstarter, July 12–August 11, 2011.
Video, 6:44. https://www.kickstarter.com/projects/1996857943/the-misadventures
-of-awkward-black-girl.

———. "Our Un-Awkward Conversation with 'Awkward Black Girl' Creator Issa Rae."
Interview by Jen Ortiz. *Paper*, August, 13, 2012. http://www.papermag.com/our-un
-awkward-conversation-with-awkward-black-girl-creator-issa-rae-1426237924.html.

Rainie, Lee, and Barry Wellman. *Networked: The New Social Operating System*. Cambridge, MA: MIT Press, 2012.

Ramey, Garey, and Valerie Ramey. "The Rug Rat Race." *Brookings Papers on Economic
Activity*, Spring 2010.

Rao, Tejal. "A New Generation of Food Magazines Think Small, and in Ink." *New York
Times*, March 27, 2018. https://www.nytimes.com/2018/03/27/dining/food
-magazines.html.

Reece, Jason. "The Detroit Neighborhood Opportunity Index." Presentation to the
Neighborhood Funders Group, Detroit, July 10, 2014.

Reese, Laura A., Jeanette Eckert, Gary Sands, and Igo Vojnovic. " 'It's Safe to Come,
We've Got Lattes': Development Disparities in Detroit." *Cities* 60, Part A (February
2017): 367–77.

Reses, Jacqueline. "Workforce Diversity at Yahoo." Yahoo, 2014. https://yahoo.tumblr
.com/post/89085398949/workforce-diversity-at-yahoo.

Rideout, Victoria, and S. Craig Watkins. *Millennials, Social Media, and Politics.* NORC,
University of Chicago; Institute for Media Innovation, University of Texas at Austin, forthcoming 2019.

Rose, Steve. "Justin Simien: 'I'm Black, I'm a Man, I'm Gay, but I'm More Than All of
Those Things.'" *Guardian*, June 25, 2015. https://www.theguardian.com/film/2015
/jun/25/justin-simien-im-black-im-a-man-im-gay-but-im-more-than-all-of-those
-things.

Roy, Jessica. "Twitter User Appears to Have Live-Tweeted the Shooting of Michael
Brown." *New York Magazine*, August 15, 2014.

Satariano, Adam, and Lucas Shaw. "SoundCloud Gets New Life with Fresh $170
Million Investment." *Bloomberg*, August 11, 2017. https://www.bloomberg.com
/news/articles/2017-08-11/soundcloud-gets-new-life-with-fresh-170-million
-investment.

Saxena, Jaya. "Activists Launch Comprehensive 'Resistance Manual' for Political Action." *Daily Dot*, January 17, 2017. https://www.dailydot.com/irl/resistance-manual.

Schement, Jorge Reina. "Wiring the Castle: Demography, Technology and the Transformation of the American Home." Lecture at MIT Communication Forum: Media
in Transition, Cambridge, MA, March, 2006.

Schreier, Jason. *Blood, Sweat, and Pixels: The Triumphant, Turbulent Stories Behind How
Video Games Are Made.* New York: Harper Collins, 2017.

Scott, Tom. "Renegade Rapper Prince Harvey Records Full Album in Apple Store."
Interview. Red Bull Music Studios, Auckland. July 18, 2015. http://www.redbull
studios.com/auckland/articles/renegade-rapper-prince-harvey-records-full-album
-in-apple-store-interview.

Scott, Veronika. "Empowerment Plan: Veronika Scott." Ponyride, 2016. https://www
.ponyride.org/resident-stories/2016/11/17/empowerment-plan.

———. "Veronika Scott: The Empowerment Plan." May 22, 2012, San Jose, CA. Tedx
video, 13:59. https://www.youtube.com/watch?v=dmqM8vZDt3I.

Shaw, Lucas. "Netflix Hires Executive to Help Deal with Its Diversity Problem." *Bloomberg*, June 26, 2018. https://www.bloomberg.com/news/articles/2018-06-26/new
-netflix-executive-to-help-make-staff-as-diverse-as-its-users.

Simien. Justin. "The Birth of *Dear White People.*" *Huffington Post*, June 14, 2012. https://
www.huffingtonpost.com/entry/dear-white-people_b_1598095.html.

———. *Dear White People*: Concept Trailer. YouTube, June 13, 2012. https://www
.youtube.com/watch?v=watjO62NrVg&feature=player_embedded.

———. "*Dear White People* Director Sundance Film Festival's 'Breakthrough Talent.'"
Interview by Hillary Sawchuk. *A Drink With*, January 30, 2014. http://adrinkwith
.com/justin-simien.

———. *Dear White People.* Indiegogo, June 15–July 14, 2012. https://www.indiegogo
.com/projects/dear-white-people#.

———. "*Dear White People* Is a Satire Addressed to Everyone." Interview by Terry
Gross. *Fresh Air*, October 16, 2014. Audio, 26:11. https://www.npr.org/programs
/fresh-air/2014/10/16/356692662/fresh-air-for-october-16-2014.

———. "How to Overcome NO & Beat the System with Justin Simien." Interview by Chase Jarvis. *Chase Jarvis LIVE*, December 10, 2014. Video, 1:25:29. https://www .youtube.com/embed/WLgKhA7dlUE.

———. "Justin Simien's Open Letter." Interview by Zack Etheart. *Interview*, January 28, 2014. https://www.interviewmagazine.com/film/justin-simien-dear-white-people.

———. "The Man Behind *Dear White People*." Interview on *Newsroom*. CNN, July 17, 2012. https://www.youtube.com/watch?v=JJuXarfrxOE. Video, 2:19.

Simon, Mashaun D. "NBCBLK28: Samuel Sinyangwe: Number Cruncher in the Fight Against Systemic Racism." NBC News Online, February 8, 2017. https://www.nbc news.com/feature/nbcblk28-2017/nbcblk28-samuel-sinyangwe-number-cruncher -fight-against-systemic-racism-n714771.

Sinyangwe, Samuel. "Mapping Police Violence." Presentation at the Data & Civil Rights Conference: A New Era of Policing and Justice, Washington, DC, October 27, 2015. Video, 8:16. http://www.datacivilrights.org/2015.

Sinyangwe, Samuel, DeRay Mckesson, and Brittany Packnett. *2015 Police Violence Report.* Mapping Police Violence, 2015. https://mappingpoliceviolence.org/2015.

Sisario, Ben. "Algorithm for Your Personal Rhythm." *New York Times*, January 11, 2014. https://www.nytimes.com/2014/01/12/arts/music/beats-music-enters-online -streaming-market.html.

———. "Drake's 'Scorpion' Stays on Top as Hip-Hop Dominates Billboard's Top 5." *New York Times*, July 16, 2018, https://www.nytimes.com/2018/07/16/arts/music /drake-scorpion-second-week-billboard-chart.html.

Sisario, Ben, and Joe Coscarelli. "XXXTentacion Signed $10 Million Album Deal Before His Death." *New York Times*, July 8, 2018, https://www.nytimes.com/2018/07/08 /arts/music/xxxtentacion-death-album.html.

Smith, Aaron, and Monica Anderson. *Social Media Use in 2018.* Pew Research Center, March 2018.

Smith, Stacy L., Marc Choueiti, Katherine Pieper, with Ariana Case, Kevin Yao, and Angel Choi. *Inequality in 900 Popular Films: Examining Portrayals of Gender, Race/ Ethnicity, LGBT, and Disability from 2007–2016.* Media, Diversity, and Social Change Initiative Annual Report. USC Annenberg, July 2017. https://annenberg.usc.edu /sites/default/files/Dr_Stacy_L_Smith-Inequality_in_900_Popular_Films.pdf.

Snyder, Thomas D., Cristobal de Brey, and Sally A. Dillow. "Bachelor's Degrees Conferred by Postsecondary Institutions, by Race/Ethnicity and Field of Study: 2012–13 and 2013–14." In *Digest of Education Statistics, 2015.* 51st ed. Table 322.30. Washington, DC: National Center for Education Statistics, 2016. https://nces.ed .gov/programs/digest/d15/tables/dt15_322.30.asp.

———. "Degrees Conferred By Postsecondary Institutions, by Level of Degree and Sex of Student: Selected Years, 1869–70 through 2026–27." In *Digest of Education Statis- tics, 2016.* 52nd ed. Table 318.10. Washington, DC: National Center for Education Statistics, 2018. https://nces.ed.gov/programs/digest/d16/tables/dt16_318.10.asp ?referrer=report.

———. "Number and Percentage Distribution of Science, Technology, Engineering, and Mathematics (STEM) Conferred by Postsecondary Institutions by Race/ Ethnicity, Level of Degree/Certificate, and Sex of Student: 2008–09 through

2013–14." *Digest of Education Statistics, 2015.* 51st ed. Table 318.45. Washington, DC: National Center for Education Statistics, 2016. https://nces.ed.gov/programs /digest/d15/tables/dt15_318.45.asp.

Snyder, Thomas D., and Sally A. Dillow. "Degrees in Computer and Information Sciences Conferred by Degree-Granting Institutions, by Level of Degree and Sex of Student: 1970–71 through 2010–11." *Digest of Education Statistics, 2012.* 48th ed. Table 349. Washington, DC: US Department of Education National Center for Education Statistics, 2013. https://nces.ed.gov/programs/digest/d12/tables/dt12_349.asp.

Sow, Aminatou. "Podcast Co-Host Helps Women Find Each Other." American Voices, *Time,* 2017. Video, 1:56. http://time.com/collection/american-voices-2017 /4978322/aminatou-sow-american-voices.

Spinuzzi, Clay. "Working Alone Together: Coworking as Emergent Collaborative Activity." *Journal of Business and Technical Communication* 26, no. 4 (October 2012): 399–441.

Spreitzer, Gretchen, Peter Bacevice, and Lyndon Garrett. "Why People Thrive in Co-working Spaces." *Harvard Business Review,* September 2015.

Stafford, Kat. "50 Detroiters: 'What Happened to the People That Were Here.'" *Detroit Free Press,* December 13, 2015.

Stein, Joel. "Millennials: The Me Me Me Generation." *Time,* May 20, 2013. http://time .com/247/millennials-the-me-me-me-generation.

Sterling, Debbie. "How the Founder of GoldieBlox Is Creating the Next Generation of Women in STEM." Interview by Elana Lyn Gross. *Forbes,* October 11, 2017. https://www.forbes.com/sites/elanagross/2017/10/11/how-the-founder-of -goldieblox-is-creating-the-next-generation-of-women-in-stem/#1270c3644e13.

———. "Yesterday's Frontiers Tomorrow's Horizons." Filmed April 19, 2013, in State College, PA. TEDxPSU video, 17:08. https://www.youtube.com/watch?v=FE eTLopLkEo.

Suarez, Gary. "Once a SoundCloud Sensation, Lil Pump Proves a Hot 100 Contender." *Forbes,* October 24, 2017. https://www.forbes.com/sites/garysuarez/2017/10/24 /lil-pump-smokepurpp-billboard/#5370e1581fd8.

Sugrue, Thomas J. *The Origins of the Urban Crisis: Race and Inequality in Postwar Detroit.* Princeton, NJ: Princeton University Press, 1996.

Sundararajan, Arun. *The Sharing Economy: The End of Employment and the Rise of Crowd-Based Capitalism.* Cambridge, MA: MIT Press, 2017.

Takeuchi, Lori, Reed Stevens, et al. *The New Coviewing: Designing for Learning Through Joint Media Engagement.* New York: Joan Ganz Cooney Center, 2011.

Tang, Eric, and Chunhui Ren. *Outlier: The Case of Austin's Declining African-American Population.* Issue Brief, Institute for Urban Poverty Research and Analysis, University of Texas at Austin, May 8, 2014. https://liberalarts.utexas.edu/iupra/ _files/pdf/Austin%20AA%20pop%20policy%20brief_FINAL.pdf.

Tate, Ashley. "The Tweet That Launched a Movement." *Nation Swell,* May 25, 2015. http://nationswell.com/tweet-launched-movement-wetheprotesters.

Taylor, Paul, Rick Fry, and Russ Oates. *The Rising Cost of Not Going to College.* Washington, DC: Pew Research Center, February, 2014.

Thompson, Derek. "Quit Your Job." *Atlantic,* November 5, 2014. https://www.the atlantic.com/business/archive/2014/11/quit-your-job/382402.

Toossi, Mitra, and Teresa L. Morisi. "Women in the Workforce Before, During, and After the Great Recession." US Bureau of Labor Statistics, Spotlight on Statistics, July 2017. https://www.bls.gov/spotlight/2017/women-in-the-workforce-before -during-and-after-the-great-recession/pdf/women-in-the-workforce-before-during -and-after-the-great-recession.pdf.

Turkle, Sherry. *Alone Together: Why We Expect More from Technology and Less from Each Other*. New York: Basic Books, 2011.

Turner, David. "Look at Me! The Noisy, Blown-Out SoundCloud Revolution Redefin- ing Rap." *Rolling Stone*, June 1, 2017. https://www.rollingstone.com/music/music -features/look-at-me-the-noisy-blown-out-soundcloud-revolution-redefining -rap-123887.

US Bureau of Labor Statistics. "Computer and Information Technology Occupations." *Occupational Outlook Handbook*. Washington, DC: Office of Occupational Statistics and Employment Projections, April 2018. https://www.bls.gov/ooh/computer-and -information-technology/home.htm.

————. "Contingent and Alternative Employment Arrangements News Release." Washington, DC: US Department of Labor, June 7, 2018. https://www.bls.gov /news.release/conemp.htm.

————. *Employment Projections—2016–2026*. Washington, DC: US Department of Labor, last modified date, January 30, 2018.

————. *Highlights of Women's Earnings in 2016*. BLS Reports. Washington, DC: US Department of Labor, August 2017.

————. "Labor Force Statistics from the Current Population Survey." Washington, DC: US Department of Labor, January 2018. https://www.bls.gov/cps/cpsaat11 .htm.

————. "Median Usual Weekly Earnings of Full-Time Wage and Salary workers, by Detailed Occupation, 2016 Annual Averages." In *Highlights of Women's Earnings in 2016*. Table 2. BLS Reports. Washington, DC: US Department of Labor, August 2017. https://www.bls.gov/opub/reports/womens-earnings/2016/home.htm.

US Census Bureau. "Detroit City, Michigan." QuickFacts. Washington, DC: US De- partment of Commerce, 2017. https://www.census.gov/quickfacts/detroitcity michigan.

————. "Millennials Outnumber Baby Boomers and Are Far More Diverse, Census Bureau Reports." Washington, DC: US Department of Commerce, June 25, 2015.

————. "Young Adults Then and Now." Infographic. Washington, DC: US Department of Commerce, accessed October 25, 2018. https://census.socialexplorer.com /young-adults/#/report/full/nation/US.

US Department of Education. National Center for Education Statistics, Integrated Postsecondary Education Data System, Fall, Completions component.

Urbain, Thomas. "Facebook as an Election Weapon, from Obama to Trump." Phys.org, March 23, 2018. https://phys.org/news/2018-03-facebook-election-weapon-obama -trump.html.

Valkenburg, Patti M., Marina Krcmar, Allerd L. Peeters, and Nies M. Marseille. "Developing a Scale to Assess Three Styles of Television Mediation: 'Instructive Mediation,' 'Restrictive Mediation,' and 'Social Coviewing.'" *Journal of Broadcasting and Electronic Media* 43, no. 1 (1999): 52–66.

Van Buskirk, Eliot. "SoundCloud Threatens MySpace as Music Destination for Twitter Era." *Wired*, July 6, 2009. https://www.wired.com/2009/07/soundcloud-threatens -myspace-as-music-destination-for-twitter-era.

van Dijk, Jan A. G. M. "The Evolution of the Digital Divide: The Digital Divide Turns to Inequality of Skills and Usage." In *Digital Enlightenment Yearbook*, edited by Jacques Bus, Malcolm Crompton, Mireille Hildebrandt, and George Metakides, 57–75. Fairfax, VA: IOS Press, 2012.

Vara, Vauhini. "Why Doesn't Silicon Valley Hire Black Coders?" *Bloomberg Businessweek*, January 21, 2016. https://www.bloomberg.com/features/2016-howard-university -coders.

Vinsel, Lee. "Design Thinking Is a Boondoggle." Chronicle Review, *Chronicle of Higher Education*, May 21, 2018. https://www.chronicle.com/article/Design-Thinking-Is -a/243472.

von Zastrow, Claus. "Does Computer in High School Face a Bright Future?" Education Commission of the States. September 23, 2016. https://www.ecs.org/does -computer-science-in-high-school-face-a-bright-future.

Warschauer, Mark. *Technology and Social Inclusion: Rethinking the Digital Divide*. Cambridge, MA: MIT Press, 2003.

Watkins, S. Craig. "The Evolution of #Black Twitter." In *Signs of Life in the USA: Readings on Popular Culture for Writers*. 9th ed. Edited by Sonia Maasik and Jack Solomon, 445–49. New York: Macmillan Press, 2018.

———. *Hip Hop Matters: Politics, Pop Culture, and the Struggle for the Soul of a Movement*. Boston: Beacon Press, 2005.

———. *Representing: Hip Hop Culture and the Production of Black Cinema*. Chicago: University of Chicago Press, 1998.

Watkins, S. Craig, with Andres Lombana-Bermudez, Alexander Cho, Vivian Shaw, Jacqueline Ryan Vickery, and Lauren Weinzimmer. *The Digital Edge: How Black and Latino Youth Navigate Digital Inequality*. New York: New York University Press, 2018.

Wei, Lu. "Number Matters: The Multimodality of Internet Use as an Indicator of the Digital Inequalities." *Journal of Computer-Mediated Communication* 17, no. 3 (April 2012): 303–18.

Wells, John, and Laurie Lewis. *Internet Access in U.S. Public Schools and Classrooms: 1994–2005*. NCES 2007–020. Washington, DC: US Department of Education, National Center for Education Statistics, 2006.

Westervelt, Eric. " 'Disrupting' Tech's Diversity Problem with a Code Camp for Girls of Color." *NPR Ed* (blog), August 17, 2015. https://www.npr.org/sections/ed/2015 /08/17/432278262/hacking-tech-s-diversity-problem-black-girls-code.

Weststar, Johanna, Victoria O'Meara, and Marie-Josée Legault. *Developer Satisfaction Survey 2017: Summary Report*. International Game Developer Association, January 8, 2017. https://cdn.ymaws.com/www.igda.org/resource/resmgr/2017_DSS_ /!IGDA_DSS_2017_SummaryReport.pdf

Williams, Maxine. "Building a More Diverse Facebook." Facebook Newsroom, June 25, 2014. https://newsroom.fb.com/news/2014/06/building-a-more-diverse -facebook.

Yahoo Finance. "Oakland Teen Develops App to Combat Police Brutality." February 18,
  2016. https://finance.yahoo.com/news/oakland-teen-develops-app-combat
  -151211019.html.
"Young Coder Finds Like Minds at Oakland Hackathon." KQED News, March 4, 2014.
  Video, 2:34. https://www.youtube.com/watch?v=5tnXsnAaEF4.
Zellner, Xander. "XXXTentacion Leads Billboard Artist 100 for First Time, Sparked by
  '?' Debut." *Billboard*, March 28, 2018. https://www.billboard.com/articles/columns
  /chart-beat/8265156/xxxtentacion-leads-billboard-artist-100.
Zubrzycki, Jackie. "Should Students Be Able to Replace Foreign Language with Cod-
  ing?" *Curriculum Matters* (blog), *Education Week*, June, 2, 2016. http://blogs.edweek
  .org/edweek/curriculum/2016/06/should_students_be_able_to_rep.html.

INDEX

accelerator model, 174–76
access gap, 142–43, 207–8
activism: about, 159–60; civic decline narrative, 161–62; civic innovation narrative, 162–64; civic start-ups, 174–76; connected activism, 163–64, 169–71, 182–83; data activism, 171–74; Ferguson shooting (Michael Brown), 165–71; networked and tech savvy, 168–71, 176–77; Resistance Manual, 177–83; Twitter and social media, 177–79; WeTheProtesters, 174–82. *See also* social engagement
Advanced Placement exams, 129
advertising, online, 127–28
affordability, workspace, 20
African Americans: Black crowd power, 98, 109–10; Black Girls Code, 131–33; Black Twitter, 156, 208; Color Creative initiative, 94–96; *Dear White People*, 97–109; in Detroit, 187–88; digital divide and, 141–43, 155–56; education and employment, 6; Henderson, Michael, 25–30; Mapping Police Violence, 173–74; *The Misadventures of Awkward Black Girl* (web series), 85–94; NAEP math performances, 114–15; online communities and, 65; Qeyno

hackathons, 150–57; racial hyper-exclusion, 128–30
African entrepreneurs, 27–28
Airbnb, 8
algorithms, 127–28; 172
alternative work arrangements, 12–13, 21
Amazon, 38, 53
"American Dream," 11, 57
Apple, 17, 18, 98, 111
Apple design lab, 17
Apple Music, 78, 79
Apple Store, 15–17
artificial intelligence, 127, 128, 145
Asian/Pacific Islander Americans, 114–15
aspirational labor, 56
Audacity, 44
Austin, TX: Chicon Collective, 23–30; Juegos Rancheros, 31–35, 40–44; migration and demographic changes, 37–38; North Door, 30–35
automation, 145
Axelrod, David, 170–71

Beats Music, 78
Beck, 72
Bennett, Cole, 77
BioWare, 37, 38
Black at Yale (film), 101
Black Girls Code, 131–33, 135–37

231

ABOUT THE AUTHOR

S. Craig Watkins is the founding director of the Institute for Media Innovation (IMI) and the incoming Ernest S. Sharpe Centennial Professor in the Moody College of Communication at the University of Texas at Austin. IMI is a boutique hub for research and design that poses cutting-edge questions about the intersections between media, technology, and innovation. An internationally recognized expert in media, Watkins is the author of five books exploring young people's engagement with technology, media, and pop culture. His research on young people's media practices has been supported by the MacArthur Foundation and featured in *Time*, the *Washington Post*, the *Atlantic*, and the *New York Times*, on National Public Radio and ESPN, and at the Aspen Institute and SXSW.